DATA OVER RADIO

Data and Digital Processing Techniques in Mobile and Cellular Radio

Geoff Varrall
Roger Belcher
RTT Systems Ltd.

QUANTUM PUBLISHING

Editor and Publisher *Paul Mandelstein*
Project Management, Electronic Composition,
and Technical Illustrations *Professional Book Center*
Design *Lee Ballentine*
Text Programming *Bruce Boston*
Proofreading *Alison Stuart*

Published by Quantum Publishing, Inc.
P.O. Box 310
Mendocino, California 95640

Library of Congress Cataloging-in-Publication Data

Varrall, Geoffrey.
Data over radio / Geoffrey Varrall, Roger Belcher.
p. cm.
Includes index.
ISBN 0-930633-14-8 : $95.00
1. Mobile communication systems. 2. Data transmission systems
3. Cellular radio. I. Belcher, Roger. II. Title.
TK6570.M6V37 1992
621.382′5--dc20 92-10174
CIP

ISBN 0-930633-14-8
Printed in the United States of America

The authors and publisher gratefully acknowledge the following entities and individuals for permission to reproduce the following figures:

Fig. 7.2: SMS, Cambridge, U.K.; Fig. 7.4: Dowty Information Systems, Chippenham, Wiltshire, U.K.; Fig. 7.5: Vodata and Orbitel, U.K.; Fig. 7.7: CDLC Modem and Cellphone; Fig. 7.10: Paknet Ltd., Thatcham, Berks., U.K.; Fig. 7.11: Cognito Group, Ltd., Newbury, Berks., U.K.; Figs 8.4, 8.11, 8.12: Wandel and Goltermann, UK; Fig. 9.1: Hewlett Packard (Test Instrumentation), UK; Fig. 9.3: Hughes (Networks) Corporation, USA; Fig. 15.1: Simplex Lighting, UK; Fig. 10.1: Plessey, Ltd., UK

CONTENTS

APPENDICES

INTRODUCTION

Our primary concern in this book is to look at ways of delivering data over mobile radio—delivering a data stream that represents text (words), numeric information (numbers), or speech (voice).

Data Over Radio also compares the data-bearing capability of today's analog systems with GSM (Pan-European Digital Cellular), ADC (American Digital Cellular), and JDC (Japan Digital Cellular).

We have strived to present an objective assessment of the technology options, the implicit complexity, cost and performance trade-offs, and provide a route map of how we see the mobile radio industry developing over the next 10 to 15 years.

Here, we will introduce some of the key concepts for understanding our subject and provide some useful background information.

SOURCE CODING

Coding of input information is known as *Source Coding*. An early example is Morse code—dots and dashes representing information, for example "SOS." Morse code was used in line-transmission-wired telegraph systems in the United States from the 1850s onward and in wireless systems from the late 1890s onward.

Twentieth century examples of source coding include Baudot coding and ASCII—the American standard for communications information interchange. Baudot code, widely used in early teleprinter applications, used a five-number binary sequence. ASCII uses a 7-bit digital sequence to represent 128 characters—letters of the alphabet, numbers, punctuation, spacing—sometimes shortened to 6-bit ASCII (64 characters) or lengthened to 8 bits with an additional parity bit. Representing speech, the complex analog audio waveform produced by the human voice, is more of a challenge.

Nyquist, working in the 1920s, proved that in theory and in practice a complex analog sinusoidal waveform could be reconstructed completely from a uniformly spaced set of discrete time samples, provided the sampling rate was equal to twice the highest speech frequency.

The most common standard in line transmission has been and still is 64 kb/s (kilobits per second) PCM—pulse code modulation; i.e., an analog waveform described at a moment in time by an 8-bit digital word at a sampling rate of 8 kHz ($8 \times 8 = 64$).

The description process, known as quantizing or quantization, produces unwanted quantization noise. The higher the number of quantization levels, the lower the noise—for example a 13-bit digital word giving a finer resolution of the input waveform will have lower quantization noise than an 8-bit quantizer but will require a higher data rate ($13 \times 8 = 104$ kb/s).

PCM can be optimized by making the process nonlinear. Voice communication channels typically have low levels of speech, interspersed with occasional speech peaks.

Nonlinear or nonuniform quantization provides fine quantization of the low-level weak signals and coarse quantization of strong signals—the "peaks"—trading off reduced noise for the weak signals against increased noise for the peaks. The European PCM standard uses an "A" law compression characteristic and hence is known as "A" Law PCM.

The higher data rates of in-line transmission and radio transmission applications imply more bandwidth. Nyquist demonstrated mathematically a maximum theoretical limit for transmitting binary data of 2 bits per second per Hertz using an ideal (i.e., impossible to implement) filter. With practical filters, the limit today is nearer 1.6 bits per second per Hertz; therefore our 64 kb/s PCM word occupies a minimum bandwidth of 40 kHz.

Mobile radio, working within tight spectrum constraints, requires a more efficient way of describing the input information.

A typical voice conversation has considerable built-in redundancy—information that can be discarded without compromising perceived speech quality at the receiver—and long periods of silence (the duty cycle). Speech also exhibits a high degree of predictability from sample to sample.

More "efficient" variants of PCM send the "difference" value from sample to sample or, more efficiently, predict the change in value from sample to sample and send the prediction difference or prediction residual, with a quantizing step size dynamically adapted to the characteristics of the input waveform—adaptive differential pulse code modulation (ADPCM). These techniques can reduce the net data rate from 64 kb/s to 32 kb/s or 16 kb/s.

Below 16 kb/s, waveform coding reaches a point of diminishing returns—the output from the coder becomes less resilient to channel errors, causing speech-quality degradation.

Lower bit rate representations of voice need to use speech synthesis. A speech synthesizer provides a numerical description of a speech sample segment, typically 20 millisecond (ms), expressing the tonal qualities of the segment by a series of digital filter coefficients, and the temporal qualities by an excitation sequence. "Toll quality" speech, more or less equivalent to PSTN (public sector telephone network) quality using today's technology, can be delivered at bit rates of 13 kb/s (GSM) or 8 kb/s (ADC).

CODING AND ACCESS OVERHEAD

These are *net* bit rates and do not take into account the error protection (detection and correction), multiplexing, and multiple access overhead. GSM's gross bit rate of 270 kb/s delivers 8 × 13 kb/s voice channels (104 kb/s) accompanied by 166 kb/s (62% of the gross bit rate) of channel coding, multiple access, and signaling overhead.

Lower bit rate codecs generally require higher coding overhead—the point of diminishing returns.

Digital cellular radio as a data bearer or delivery mechanism for digitized speech requires a complex processing environment but promises, in the long run, lower infrastructure cost, lower terminal costs, better spectrum utilization (more users per MHz of spectrum), lower power budgets for handheld terminals (better talk-time-to-size-to-weight ratio), and a wider choice of voice, text, and data services than existing analog systems offer.

ACKNOWLEDGMENTS

Much of the factual information presented in *Data Over Radio* has been compiled during seven years of preparing technology seminars in the United Kingdom, Europe, and elsewhere. We would like to thank our co-presenters: Mike Fitch (British Telecom Research); David Ogley (SkyMast Communication); John Davies; John Doble; Dr. Hamid Aghvami (King's College, London); Dr. Ray Macario (University of Swansea); Dr. Chris Gibbins (Rutherford Appleton Laboratory); Don Baker, Aerial

Sites PLC, George Skorkowski (Digital Signal Processing Group); and Dr. Herschel Shosteck (Herschel Shosteck Associates, Washington DC).

Our thanks also to the network providers and manufacturers (chip foundries; infrastructure, terminal. and test equipment manufacturers) who have provided much valuable information. Our special thanks to Vodafone, Cellnet, and Marconi Instruments in the United Kingdom.

Last, but not least, our thanks to Lorraine Gannon for her heroic work on the manuscript.

To prosper in the longer term, the mobile radio industry must continue to deliver "appropriate" technical solutions that effectively meet clearly defined user needs. We hope *Data Over Radio* demonstrates that the industry is moving in a positive direction toward a digital future that will deliver real benefits to the industry *and* the user.

Geoff Varrall
Roger Belcher
London

CHAPTER

1

DEVICES AND TECHNOLOGY

There have been two significant "quantum leaps" in device-enabling technology since the Second World War—first, transistors and the associated production technology of printed circuit boards in the 1950s and 1960s, and second, integrated circuit technology, device miniaturization and surface-mount techniques in the 1970s and 1980s.

Improvements in device technologies must be measured against the objectives of:

Power efficiency

Lower power requirements for hand-held terminals, giving a better ratio of talk-time to weight to size.

Spectral efficiency

More users per MHz of spectrum.

Cost

of infrastructure, switches, base stations, and mobiles.

In terms of rf performance, we are looking for improvements in stability (over temperature and time), sensitivity and selectivity—improved performance in conditions of adjacent channel or co-channel interference—resilience against spurious or intermodulation products, resilience against blocking and desensitization.

PASSIVE DEVICES

Developments in passive components have included new dielectric materials—plastics, ceramics, titanium electrolytes—improved ferrites (higher stability inductors), new helical resonator devices, SAW (surface acoustic wave) devices (giving significant improvements in tuned circuits), thin/thick film technology, and new display technology. As in most other fields of electronic component production, accuracy of process has improved dramatically. This allows much closer specification and tighter tolerances especially in such areas as filters and resonators.

ACTIVE DEVICES

Silicon

The major advances in silicon technology have related principally to larger wafer capacities, coupled with higher yields, resulting in lower production costs and a scaling down of geometries resulting in higher operating speeds.

Current capabilities are producing transistors with fTs in excess of 7 GHz using emitter widths of 1 to 1.5 um. Processes using emitter widths of 0.6 μm will result in fTs of 14 GHz or greater. Processes using silicon also include emitter-coupled logic (ECL), CMOS, and mixed-signal bipolar technology.

New classes of silicon components have become cost-effective in many applications—a typical example is the silicon monolithic microwave integrated circuit (MMIC). Specifications from manufacturers include devices with fTs in excess of 6 GHz, gains of 30 dB, output capability of +15 dBm, and unconditional stability. Silicon bipolar MMICs have close uniformity of specification and ease of interfacing as advantages over gallium arsenide (GaAs) devices, and (to date) increased packing density and lower cost.

GaAs—Gallium Arsenide

The traditional material and technology for high-density integration has been silicon. Technologies such as ECL for speed and CMOS for very large-scale integration (VLSI) has been the automatic choice of the system and circuit designer. GaAs was a "fringe" technology offering discrete transistors with a very low noise figure and high fTs. The requirements of applications such as satellite television and high-level telecommunications have broadened the technology field. Intense integration design work has bought GaAs forward to the point where it is now the optimal choice for many requirements.

The key to GaAs performance lies in the parameter of electron mobility. Improved mobility of up to six times is possible in N-doped material. This dramatically shorter transit time gives a much higher fT and lower internal power dissipation—key in minimizing device noise. Newer configurations of GaAs devices are making possible higher levels of integration including ASICs. Other processes to consider include enhanced material devices such as MESFETs and HEMTs. GaAs digital ICs would seem to promise fast, efficient, and powerful computing power for the future.

Geometries

The geometry of active devices is a prime factor determining the frequency response and speed of an integrated circuit. The closer the active elements are, the less distance electrons have to travel, so the faster the transit times. The narrow interconnects allow higher packing densities and also have less capacitance and inductance. Current geometries in production include processes at 0.4 to 0.5 μm.

CMOS

Within the total communication system, there are requirements to process frequencies ranging from 900 MHz or 1.8 GHz down to audio at 300 Hz and even to 50 Hz for sub-audio signaling. Until recently there was little choice in technology types for each area of application, but now technology boundaries are blurring.

CMOS (complementary metal oxide semiconductor) has the benefit of very low power dissipation, small geometries, and hence high packing densities. Until recently its limitation was speed but reduced geometries have removed this barrier in many applications.

The original 4000 series CMOS was introduced in the late 1960s and provided a low-power but slow-speed alternative to TTL (transistor-transistor logic). In the early 1980s CMOS developed further to give HCMOS (5 V high-speed CMOS) for interface to CMOS families and HCT to interface with TTL families. Further advances increased speed to produce ACT (advanced CMOS) and FCT (fast CMOS). Both these families used sub 2 micron technology. Newer developments include improvement in noise margin performance, and improvements in speed, e.g., bi-CMOS families.

ECL

ECL (emitter-coupled logic) is a mature technology that is benefiting from modern advances in processing techniques. It has always been the choice (or only option) for very high-speed signal processing. Through the 1960s, 1970s and 1980s speeds progressed from 10s of MHz to 500/600 MHz. Late 1980s–early 1990s have seen ECL manufactured with 1 and sub 1 micron geometries resulting in speeds up to 3 to 4 GHz. The technology is still relatively power hungry, and its use in mobile communications, especially in battery operated equipment, is kept to a minimum. ECL devices are invariably the choice for the dual modulus prescaler devices used in frequency synthesis applications.

DESIGN AND DEVELOPMENT TECHNIQUES

Time to Market

There has been a "sea change" in the last decade in the approach to getting a product to market ahead of competition. Companies have become aware of the considerable benefits of being first into the market in order to maximize returns from each "window of opportunity."

To achieve a reduced time to market requires a very considerable change in the fundamental approach companies take to their total product engineering program. Several "tools" or methods are available, and adoption of one or more of these should be considered with regard to a company's present engineering strengths and weaknesses. Consideration should also be given to subcontractor capabilities if parts of the product development cycle are out-of-house. It will certainly be necessary to involve both suppliers and subcontractors in this change of engineering philosophy.

Current techniques include such "tools" as:

1. **Concurrent engineering.** This is estimated to save between 40% to 50% of development costs, 20% to 30% of manufacturing costs, and reduce engineering rework considerably. The underlying principle of concurrent engineering seeks to overlap, or make parallel, tasks in the product development cycle. The objective is to replace the traditional design-test-modify, design-test-modify cycle (sequential engineering) with an output at each stage that is fully engineered (technical, production, or financial) so the rework loops are eliminated.
2. **Technology interception.** The design of products using devices not yet realized in production, i.e., designing by computer simulation. There is also a school of thought that says a company should give its design engineers the task of solving "the most difficult problem" first. It is then (supposedly) easier to solve the "simpler," intermediate design problems, i.e., "fast track" or "safe track" development options. (See Figure 1.1)

   For example, the "hardest" design challenge in Europe today would be to develop a microportable cell phone at 1.8 GHz. GSM product at 900 MHz represents (in design terms) a simpler design task.

NOTE There are counter arguments to this. The baseband processing design challenge for GSM and PCN (personal communication network) is essentially the same. The rf processing task is arguably more onerous for PCN at 1.8 GHz, but rf transmit powers (both on the up-link and down-link) are lower. Terminal designers prefer to have networks available to alpha- and beta-test their products—influencing designers to focus on GSM.

Design Optimization

The usual choice to be made is between using a standard device or architecture with a consequent compromise in "suitability for task" and a custom solution to a problem with the conse-

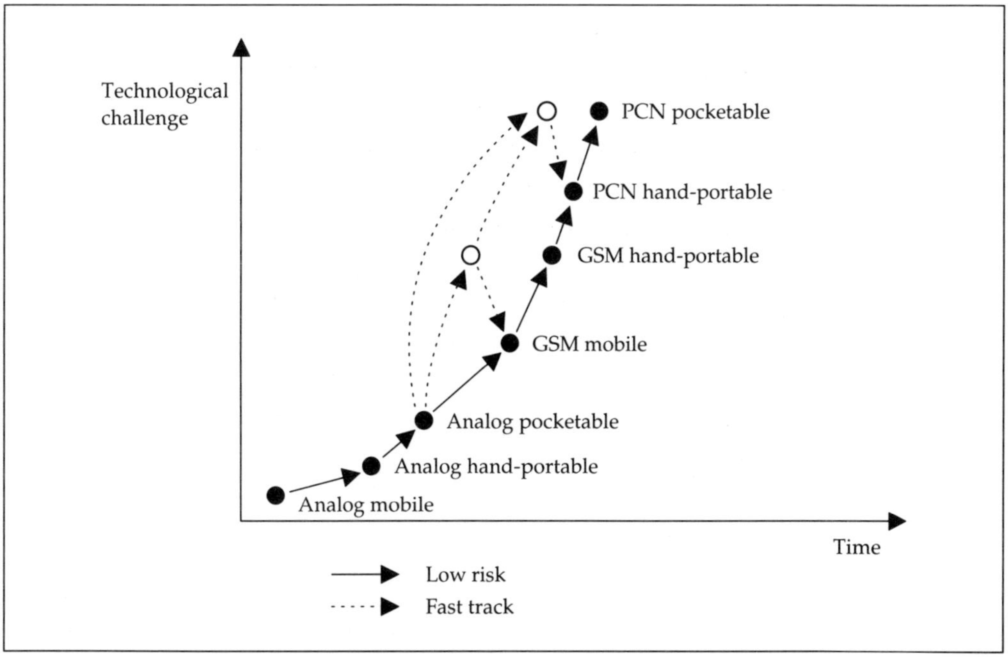

Figure 1.1 *The development routemap*

quent time scales and costs. For example, in GSM there have been a number of trade-off decisions to take: a choice of using standard digital signal processing (DSP) devices to reduce development cost, risks, and time scales or using a custom ASIC, optimized for a specific design of mobile—in theory, the bigger the forward risk, the better the long-term payoff provided the ASIC delivers.

DIGITAL PROCESSING

At the start, integrated circuit (IC) technology was divided into two very separate areas: digital circuits and analog circuits. Previous discrete component technology had also developed along the two separate streams with very little convergence of applications. Information to be processed was either analog or digital and was processed as such. If, for instance information was obtained from a transducer (analog) the signal would be amplified, filtered, scaled, and detected using discrete components, linear circuitry, and analog design techniques. Digital applications were largely confined to number processing requirements; e.g., event counting, process control requirements, etc. Digital circuits were constructed using discrete components—transistors, diodes, resistors, and capacitors.

As the capability of IC design grew, it maintained this separation of applications. The IC designer's aim was simply to replace as much of the discrete circuitry in the existing application as possible. There was little or no thought to change the way in which the function was achieved. The capability of IC technology grew, and the applications taken by transistors, diodes, etc. decreased. The IC manufacturers' ability to produce larger and larger slices of silicon and the accuracy of diffusion improved. This allowed more of the circuit function to be integrated, and the tighter tolerances and greater frequency response of devices increased the applications. Hybrid techniques were evolved that allowed IC interconnection methods to be applied to both chips and discrete components. Improved ceramic materials (substrate and packaging), coupled with these intermetallic connection processes, allowed the use of hybrid functions to be produced.

The circuit designer worked closely with the IC designer to develop new circuit techniques that maximized the benefits of IC technology. Although absolute values on an IC were still not of the required order of accuracy for many applications the intermatching ability of values was good. This meant that circuit designs changed, and many applications were seen of techniques such as balanced and double-balanced circuits.

On the digital side of electronics, progress was made to increase the types of logic functions on a chip. From simple gates, designs progressed to flip-flops and multiplexers, to counters and shift registers, to latches and data selectors. A large step forward was made with the first IC memories, both the programmable-read-only memory (PROM) and the random-access memory (RAM). These memories were used at first to replace diode matrices and small magnetic core memories.

As memory capacity and speed grew, and the number and variety of logic chips grew, the price started to move down rapidly. This increasing cost-effectiveness meant that possible digital solutions came to the notice of system engineers and the more innovative analog engineers.

Choices of solution to problems now became possible with a consideration of either a digital or an analog approach. The number and types of interface ICs increased, and this made a mixed analog and digital answer more practical. Development work progressed on digital-to-analog converters (DACs) and analog-to-digital converters (ADCs), making conversion of transducer signals into digital format possible and increasing the options for methods of display. Up until this point, the motive for the design of any particular digital IC had been to answer a digital processing problem. As the processing power of the ICs increased, the design motives broadened to address— directly—particular analog problems.

The role of electronics in general was expanding, replacing many previous mechanical functions; e.g., cash registers, measuring instruments, and domestic controls. Many of the applications demanded lower power consumption and so different integration technologies were explored; for example, CMOS and current injection logic (I^2L). Higher speeds were required as more data was processed and so faster technologies were developed; for example, TTL, Schottky TTL, ECL. Speed and power consumption were directly related to the size of the geometries on the silicon substrate and to the interconnection processes.

In the early 1970s, the microprocessor was developed. The aim was a "computer on a chip." This development gave "intelligence" and "decision" to the digital capability. It also meant nonspecific function ICs could be made and used in high volumes with the designer/engineer deciding the application of the device. As the costs came down, applications increased and a much-sought-after resource was the "software engineer." Choices of solution now encompassed analog or digital approaches, and if digital, the ratio of hardware in a solution to software (giving increased application and flexibility) became a decision factor.

The mobile communications environment requires an extensive repertoire of signal processing techniques. dc to rf amplification, adaptive filtering, multiplexing, encoding and decoding, and many other arithmetic functions. The recent development of digital signal processors and mixed-signal ASICs give rise to types of digital IC technology well suited to many of these processes.

The inclusion of large software capacities in these devices ensure that a specific silicon (or GaAs) device has application in many different areas of the communication system.

The processing complexity of digital cellular is illustrated by the increase in software content (from 60 KB in an analog phone to 500 KB in GSM) and the increase in hardware requirements (from three VLSI 5,000-gate circuits in an analog terminal to six VLSI 50,000-gate circuits in GSM) and in the amount of development effort needed—less than 40 worker-years to develop a new analog phone, more than 180 worker-years to develop a digital phone.

To summarize, it is conjectured that it will take two to three years, i.e., three generations, before digital cellular equals analog cellular hand-held size-to-weight-to-talk-time ratios. The Motorola Microtac Lite analog phone, launched in August 1991 weighed-in at 221 g, with Mars Bar dimensions, 45 minute talk-time and eight hours of standby. This performance package is achieved by using a hat box of "technology tricks" —integrated antenna and receiver rf filters, large-scale integration of the IF amplifier and modulator/demodulator, miniaturized duplex filter using high permitivity ceramics, miniaturized SAW filters, ASIC integration of the logic functions, discontinuous transmission (DTX), electronic component thicknesses less than 4 mm and 0.5 mm pitch LSI packaging.

BUS ARCHITECTURE

Parallel Bus Versus Serial Bus

Early integrated circuit device applications in mobile radio tended to be a compromise—using parallel bus architectures optimized for computer applications resulted in complex interconnections. Complex interconnections required expensive multilayer boards. The multilayer boards became the largest single cost element in the radio.

With the maturing of the mobile radio market (bigger numbers), specialized chip applications have become economical. The use of serial bus architectures has resulted in reduced interconnections and allowed the use of simple low-cost boards.

BATTERIES

Analog cellular phones, digital cellular phones, and digital cordless phones all await dramatic improvements in battery technology.

The energy density of Ni-cad batteries (approx. 38 Wh/kg) can easily be exceeded by the newer mixes of nickel metal hydride or nickel cobalt or rechargeable lithium but few, if any, of the new options have been shown to share the resilience of Ni-cad batteries in the face of persistent abuse. Ni-cads can survive up to 5000 charge cycles over a relatively large temperature gradient, with well-defined power output characteristics. Rechargeable lithium for example, will deliver higher densities (greater than 50 Wh/kg) but with a shorter recharge cycle life expectancy.

Maintenance Regimens and Batteries

Environmental regulation in Europe[1] will require all products using batteries containing cadmium to be returned to point-of-purchase or source manufacturer at the product's end of life for reprocessing or environmentally acceptable disposal, with similar regulations likely for lithium or nickel cobalt. Nickel metal hydride may provide an environmentally acceptable alternative but at a premium price.

Battery Size

Recent developments to reduce battery size/volume include techniques to maximize surface area exposed to electrolyte—for example foam techniques and plastic-bonded electrodes, which reduce dead volume within the battery shell.

Table 1.1 gives a typical size and weight comparison of Ni-cad and lithium. Note also lithium's other advantage of longer shelf life.

Research work in the United Kingdom (UK Atomic Energy Authority) is currently focused on stabilizing lithium in a plastic membrane, then using it to form the outer case of the

Table 1.1 *Sealed Ni-Cad/rechargeable lithium AA size cell comparison*

	Ni-cad	**Rechargeable lithium**
Voltage	2.4 V	2.4/3 V
Capacity	700 mA	700 mA
Available capacity after three months of self-discharge	30%	97%
Weight	$25 \times 2 = 50$ g	17 g
Volume	$7.6 \times 2 = 15.2$ cm^3	7.6 cm^3

[1] This EEC directive applies to batteries containing more than 0.025% of mercury, 0.04% of lead, or 0.0025% of cadmium.

mobile radio—i.e., the case material itself would be the battery. At the end of the battery's life, the radio is recycled. (This is unlikely to be a realistic proposition within the next 10 to 15 years).

ENCLOSURE MATERIALS

The enclosure material of the radio communication system equipment has many of the traditional tasks to fulfill and others related specifically to the radio/electronic environment.

The enclosure material must of course be strong enough for the purpose, light enough for portability, meet the cost projections, and be cosmetically pleasing. For the communications industry, the enclosure also needs to have a shield capability against electrostatic fields—both external and internal. This last requirement is achieved either by spraying on a shielding coat, e.g., nickel, after enclosure manufacture or by mixing in a suitable material at the material composition stage. The shielding regulations are becoming rigorously applied by the various authorities, and manufacturers must be aware of these to avoid market restrictions.

PRODUCT LIFE CYCLE

Dizzy targets for new product introductions are becoming the norm—a new generation of analog phones every 18 months, a new generation of digital phones every 12 months, assuming digital will catch up and overtake its analog cousin by digital generation 3—an introduction time scale that requires a close and effective partnership between the chip foundry and mobile manufacturing communities.

Meantime, arguments rage as to which device technology will deliver the best results within the tightest time and cost constraints.

CHAPTER

2

THE MARKET DYNAMIC

Huge sums of money are being spent on digital mobile radio technology on the basis that digital radio will provide a more cost-effective mechanism for supporting large numbers of users. We need some reassurance that this large pool of potential new users (a) exists, and (b) can be persuaded to buy a digital cellular phone and digital air time, in preference to the analog alternative.

We are in an era of technology change driven by advances in communications technology. In the same way that steam, railways, electricity, oil and the telephone itself transformed the lives of our forefathers, so the mobile phone is transforming the ways in which we live and do business today.

REVENUE POTENTIAL

In 1991, the telecoms industry worldwide had 600 million subscribers generating $600[1] billion of revenue. Mobile radio (excluding domestic cordless) represented 10% of this installed base with 60 million subscribers worldwide.

By the year 2000, it is estimated there will be 1000 million telecomms subscribers generating 1000 billion dollars of revenue, with telecoms displacing insurance as the world's largest industry. Twenty-five percent of the total will be mobile; i.e., 250 million subscribers. Since we know there is substantial "latent" demand for communication—for example there are 16 million Russians on the waiting list for a telephone—the numbers become believable.

Mobile radio has the benefit of being a premium tariff service. Vodafone in the United Kingdom in 1989–1990 achieved the highest revenue per subscriber for any cellular network in the world (£1,000 revenue and £500 profit per subscriber) from a subscriber base of 500,000 delivering £500 million of revenue and £250 million profit and a company market capitalization of £3 billion.

CAPACITY CONCENTRATION

Analog cellular networks are demonstrating that they can support large user communities with high nodal points of user concentration. In the United Kingdom in the later 1980s, 30% of the national mobile user population was concentrated in London, 10% of that 30% was concentrated within 10 miles of Charing Cross. Cellnet's analog TACS (total access communications system) United Kingdom network in 1991 totaled 27 switches, 800 base stations delivering 30,000 voice channels nationwide, with extensive deployment of microcells (cell radii < 500 m) in urban centres (6 microcells along the length of Oxford Street). In prac-

[1] All "dollar" amounts refer to U.S. dollars.

tice, the networks can comfortably support an average of 20 users per channel (giving an acceptable minimum of blocked and dropped cells), or 600,000 subscribers—i.e., both networks together can support 1.2 million users on the 50 MHz TACS United Kingdom analog spectrum allocation.

THE MARKET FOR DIGITAL CELLULAR

The success of digital cellular will depend on a substantial increase in usage over and above this level—a target of 20 million GSM users in Europe on 50 MHz of spectrum, 40 million PCN users on 150 MHz of spectrum at 1.8 GHz. Worldwide the expectation by the end of the decade is a mobile user base of 250 million subscribers.

For mobile radio manufacturers this implies an annual market of significant size—over 40 million new users per year—a replacement market of over 125 million users, nearly 175 million mobiles per year. (See Table 2.1)

Mobile radio manufacturers—Motorola, Oki, Mitsubishi, Nokia, Matsushita, Uniden, Sony, Toshiba, Samsung, Hitachi, Gold Star—have a vested interest in joining the race for mobile radio market share.

What percentage of this market will be digital by the end of the 1990s remains open to question.[2] Analog networks are being installed today across the world—the TACS system implemented in Italy in 1990 for example—the life expectancy of these networks could be 15 to 20 years. It is a certainty that by the late 1990s we will have a mixed analog and digital environment with analog and digital networks living side-by-side.

Against this background, let us take a detailed look at the digital processing techniques used in and planned for use in the digital mobile radio networks of the 1990s.

[2] For a robust view of the market and fiscal implications of the analog-to-digital transition, see the series of PCN and Microcellular studies by Herschel Shosteck Associates, Washington DC.

Table 2.1 *Worldwide market for mobiles**

Year	(1) Number of users (in millions)	(2) New mobiles (in millions)	(3) + Replacement = Total market 5 years prior (in millions)	(4) + 10% churn annually = 10% of current market (in millions)	Total number of mobiles supplied to market (2 + 3 + 4) (in millions)
1991	60	+10	+20	+6	36
1992	72	+12	+25	+7	44
1993	86	+14	+32	+8	54
1994	103	+17	+40	+10	67
1995	124	+21	+49	+12	82
1996	149	+25	+60	+14	99
1997	179	+30	+72	+17	119
1998	215	+36	+86	+20	142
1999	258	+43	+103	+25	171

*Assumes (conservative) 5 year average market replacement cycle; 10% annual churn; and net subscriber growth of 20%

CHAPTER

3

BASEBAND PROCESSING

DEFINITIONS

Our introduction stated simply that data over radio covers all applications where we need or want to deliver a digital bit stream of 0s and 1s over a radio channel (for our stated area of interest, a mobile radio channel).

A starting point is to look at ways in which we can handle and manipulate this data stream prior to up-converting to a radio frequency and modulation of the information to a radio carrier. Our baseband in mobile radio could be anything from a few bits per second (telemetry and telecommand applications) to several hundred kilobits per second for digital cellular.

TWO-LEVEL AND FOUR-LEVEL BASEBAND OPTIONS

In theory, our baseband information can be expressed by a variety of signal shapes and two (binary) or four (quaternary) polar or unipolar signal levels. (See Figure 3.1) Polar signals will tend to give a better performance for a given signal-to-noise ratio (approx. 3 dB benefit) over unipolar.

Four-level offers us twice the bit rate of a two-level signal for every symbol change; i.e., four-level requires half the channel bandwidth of a binary signal with the same bit rate. For example, every change of symbol in two-level represents:

0 or 1

Every change of symbol in four-level represents:

	0 0	
or	0 1	i.e., twice the information rate per symbol
or	1 1	change
or	1 0	

Every change of symbol in eight-level represents:

	0 0 0	
or	0 0 1	i.e., three times the information rate per
or	0 1 1	symbol change.
or	1 1 1	
or	1 1 0	
or	0 1 0	
or	1 0 1	
or	1 0 0	

NOTE The number of symbol changes is normally referred to as the *baud rate*; i.e., baud rate = symbol rate.

A four-level baseband signal will, however, need a four-level modulation process to transfer the information to an rf carrier (eight-level will need an eight-level modulation pro-

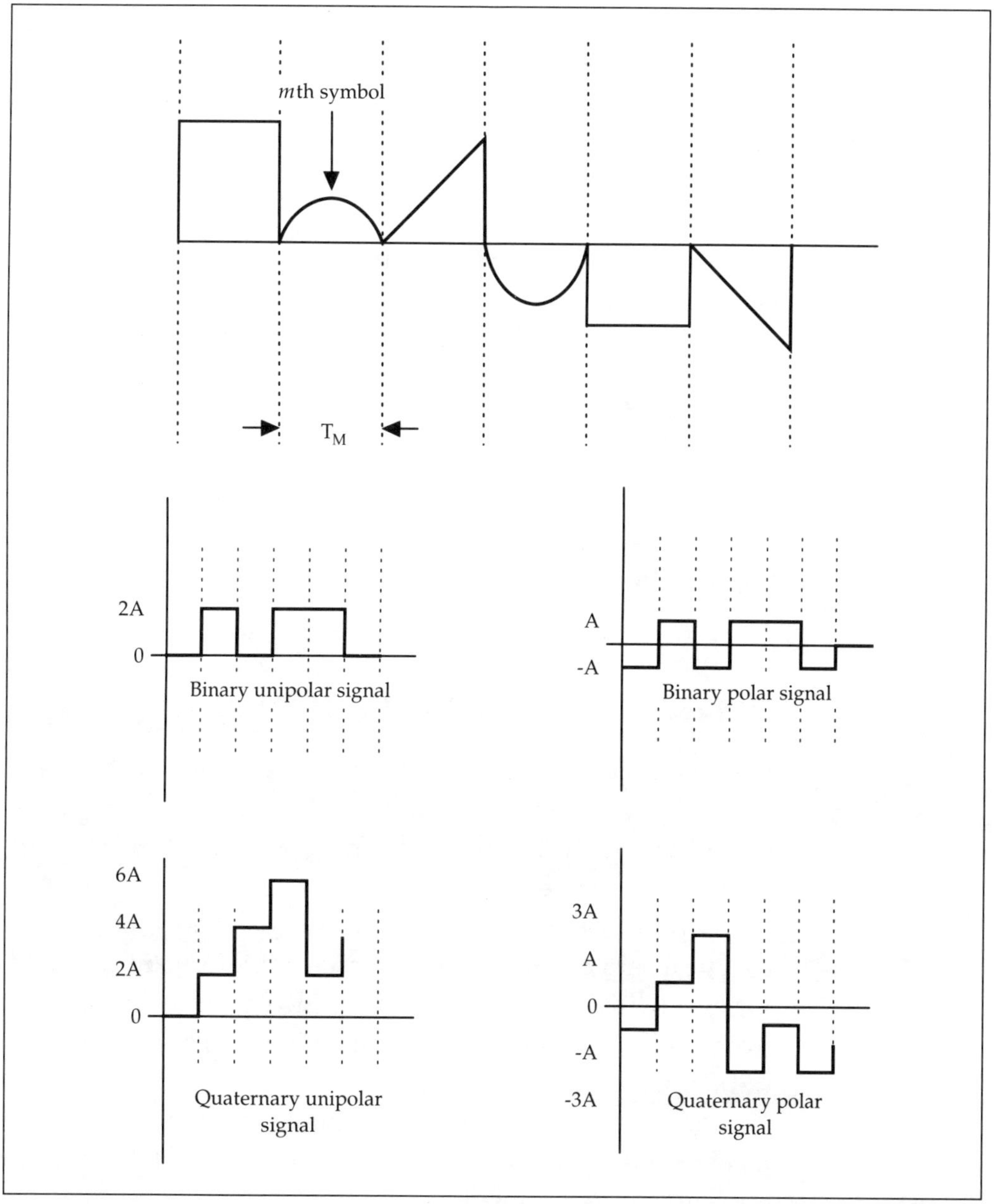

Figure 3.1 *Baseband signal: A signal containing symbols that differ in shape or in size.*

cess); for example, quarternary phase-shift keying (QPSK), eight-level frequency shift key (FSK) or eight-level phase shift key (PSK).

As we increase the number of modulation levels, we decrease the difference or distance in phase and frequency from symbol to symbol and become progressively more reliant on maintaining phase and frequency accuracy and integrity across the radio path—difficult to realize in severe multipath conditions.

Four-level modulation methods are generally the highest level practically achievable in mobile radio applications.

Rectangular pulse shaping would, in theory, be the preferred baseband description mechanism but again, prior to modulation, we will need to "soften" the transition from symbol state to symbol state. Abrupt transitions expressed as abrupt rf frequency or phase transitions cause spectral spreading if passed through nonlinear elements in the rf processing (e.g., PA [power amplification] or filter) chains.

FILTERING AND BANDWIDTH LIMITATIONS

We may therefore decide to deliberately degrade the quality of our baseband information by filtering the clean rectangular pulse, making each pulse more sinusoidal; e.g., by implementing a Gaussian filter.

Working within tight bandwidth constraints may in any event cause problems with our baseband information. Reducing bandwidth has the effect of reducing pulse width—tails from prior time slots create additive intersymbol interference (ISI). (See Figure 3.2)

We have already talked of Nyquist's minimum bandwidth theorem giving us a maximum theoretical limit for ISI-free transmission of 2 symbols/s/Hz for binary transmission using an (impossible to realize) Nyquist filter. A practical Nyquist filter, known as a raised cosine, has a predefined roll-off factor with a practical limit (today) of 0.2.

This gives us a spectral efficiency benchmark of

$$\frac{2}{1.2} = 1.66 \text{ bps/Hz}$$

As suggested above, we may already have deliberately decided to introduce some baseband intersymbol interference to smooth our symbol transitions, but we now need some benchmark to measure the amount of ISI and its related effect on our power budget.

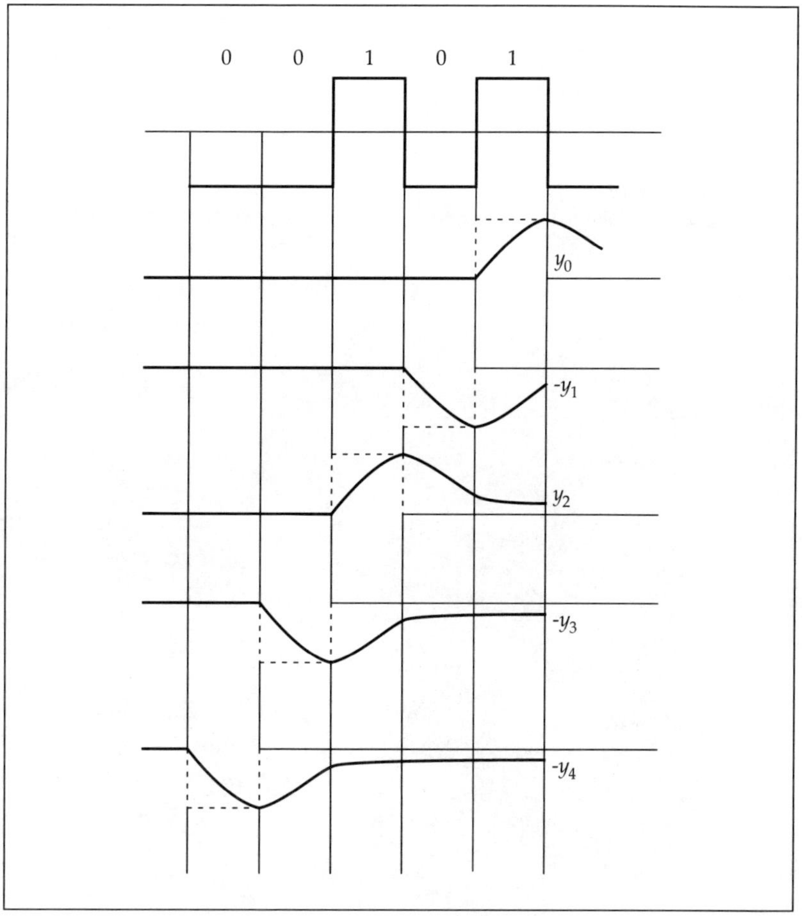

Figure 3.2 *ISI caused by the additive effect of the tails of the first, second, third, and fourth time slots.*

EVALUATING ISI

ISI caused by channel filtering imperfections is evaluated by means of an eye diagram (See Figure 3.3). An eye diagram is generated by connecting a band-limited signal to an oscilloscope triggered by the symbol clock. The time base of the oscilloscope is set to the symbol duration of the signal. The eye diagram shows the superimposition of all possible levels of the signal during the symbol duration.

A signal with no ISI distortion has a degradation of 0 dB. If the eye is only 70% open at the sampling instant, degradation due to ISI is 3 dB. Therefore ISI-free transmission needs a 3 dB increase in signal level; i.e., we have traded a decrease in power efficiency for an increase in spectral efficiency.

NOTE An overlapping signal pattern will not cross the horizontal axis at exact integer multiples of the signal rate. Deviation from the nominal crossing points known as peak-to-peak signal transition jitter can affect symbol timing recovery circuits.

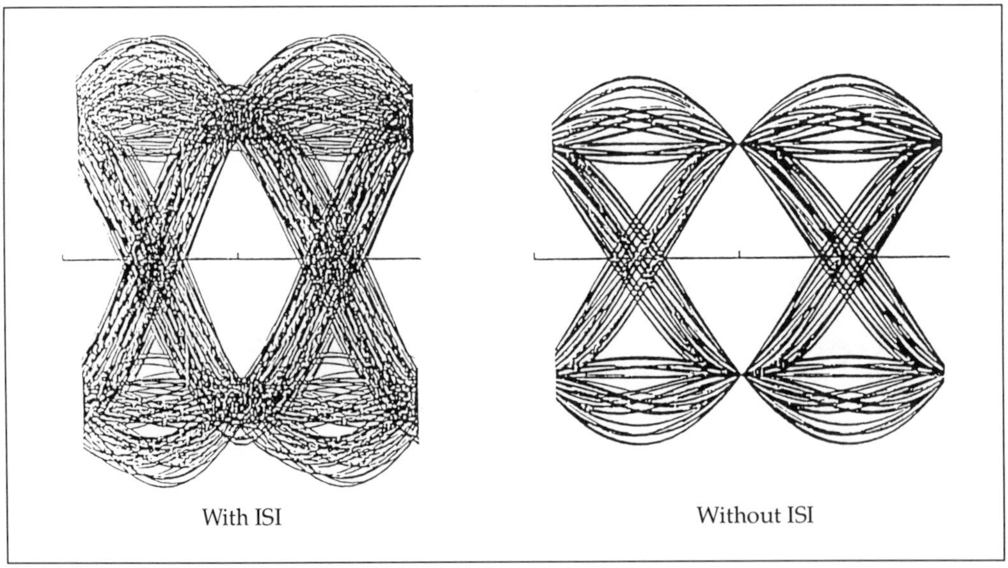

Figure 3.3

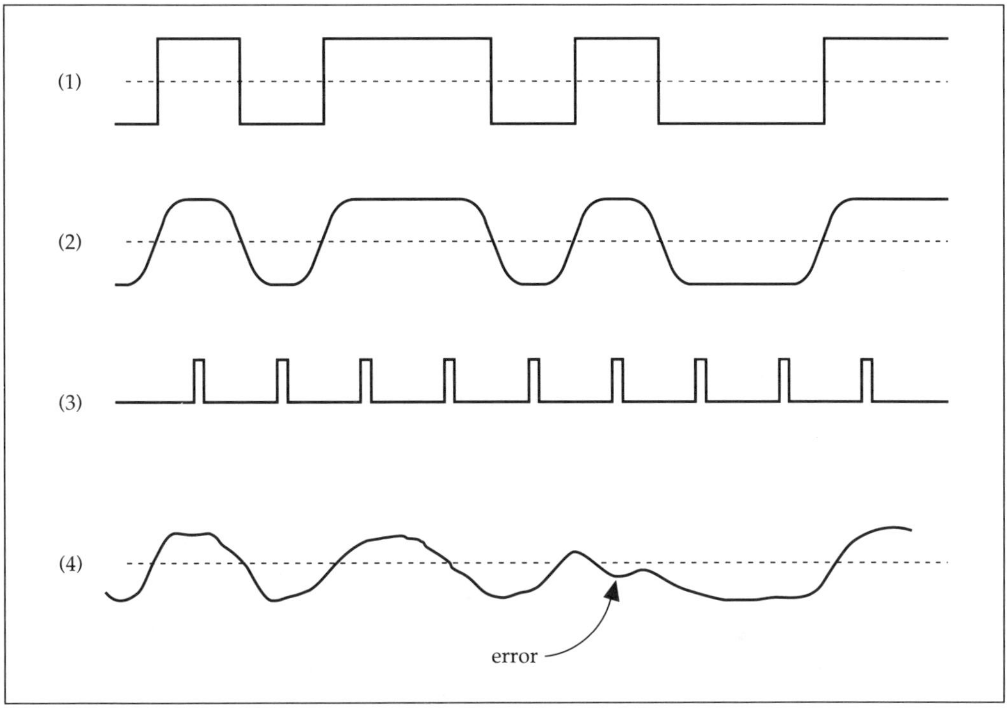

Figure 3.4

EFFECT OF WHITE GAUSSIAN NOISE

A radio transmission system will have a noise floor—an addition of the noise products (e.g., shot noise) of active devices, noise from passive devices, and the noise characteristics of the transmission channel. Statistically the noise exhibits a Gaussian probability density function and is therefore known as *additive white Gaussian noise* (AWGN). Figure 3.4 shows (1) a clear rectangular pulse; (2) filtered to soften the symbol changes; (3) with information recovery at the sampling instant; and (4) a false sampling of the information due to AWGN.

CHAPTER

4

MODULATION—THEORY AND PRACTICE

INTRODUCTION

Digital modulation is the process by which digital data is transfered to a carrier (sinusoidal waveform) in order to be transmited through a band-pass channel (radio or cable). Three parameters of the carrier can be affected to transmit information; amplitude, phase, frequency, or combinations of these. The power and bandwidth required by the carrier will be a function of modulation type or types and also the depth of modulation employed. Digital transmission uses a class of modulation referred to as *shift key modulation*. The principle of this is that the bit stream to be transmitted modifies one or more of the carrier parameters. This modification may be direct or indirect. If the phase of the carrier is used to convey information, the method is known as *coherent modulation*; if the carrier phase does not convey information, it is referred to as *noncoherent modulation*. Among coherent modulations are phase shift key (PSK), frequency shift key (FSK), amplitude shift key (ASK), and other forms of continuous phase modulation. In the noncoherent class are differential phase shift key (DPSK), and types of FSK/ASK used in a differential or nonabsolute time referenced mode. It is useful to consider the bandwidth required by such classes of modulation. To put the examination in context we will consider also basic AM and FM modulation.

From the applications point of view, digital modulation schemes may be classified into two main categories: power-efficient and spectrally (bandwidth) efficient techniques. In what follows, we assume that, for optimum performance, the overall system filtering strategy satisfies both the Nyquist (intersymbol interference free transmission) and matched filter (maximum signal-to-noise ratio at the sampling instant) criteria; i.e., having a raised-cosine roll-off shaped response equally split between modulator and demodulator. For rectangular pulse transmission, an amplitude equalizer is also assumed to be employed at the input of the "spectral shaping" filters of modulators.

ANALOG AND DIGITAL COMPARISONS

Analog Modulation

The amount of bandwidth required to transmit a modulated signal depends on the type of modulation, the depth of modulation, and the speed of modulation.

AM Systems

The signal is generated by multiplying the carrier and message signals, and adding a fraction of the result to the carrier. The AM signal formula is

$$v(t) = \text{Cos}\ \omega_c t\ (1 + m\ \text{Cos}\ \omega_m t)$$

where

m = modulation index

$\omega = 2 \times \pi \times f$

Expanded, this becomes

$$v(t) = \text{Cos } \omega_c t + 0.5m \, [\text{Cos } (\omega_c + \omega_m)t + \text{Cos}(\omega_c - \omega_m)t]$$

The time-domain and frequency-domain representation is given in Figure 4.1. Note that:

$$m = \frac{A - B}{A + B} \text{ and in this case, } \frac{A + B}{2} = 1$$

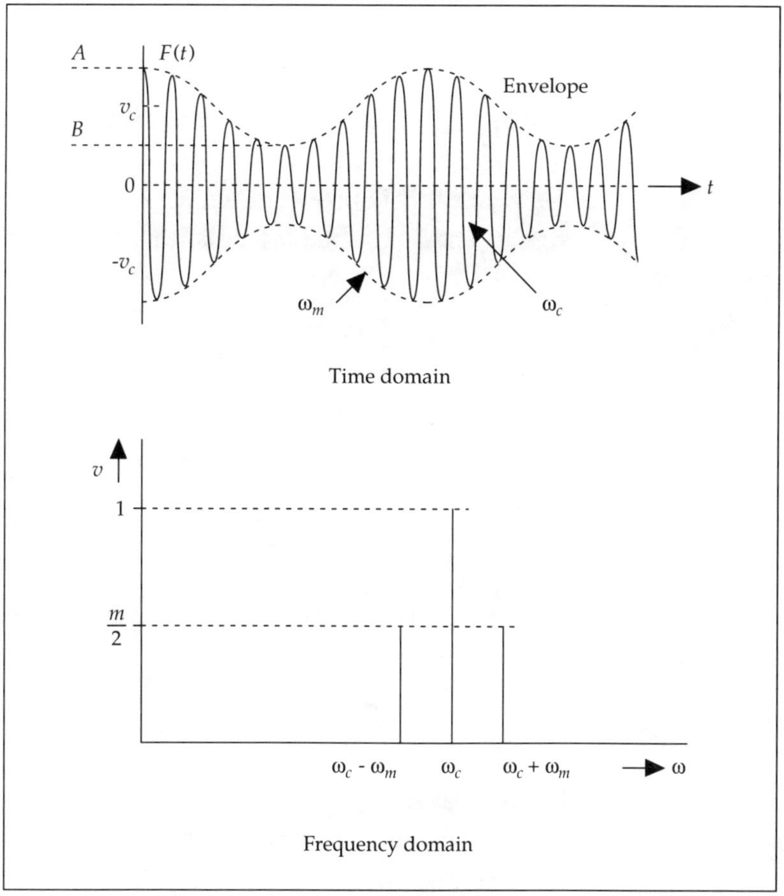

Figure 4.1

The bandwidth is $2\times\ \omega_m$ in rads/s or

$$\frac{2\times\omega_m}{2\times\pi}\ \text{Hz}$$

In practice, the message signal will not be a single frequency, but a complex pattern over a defined frequency range. If the single frequency is replaced by a complex frequency signal, then the modulated bandwidth is:

$$BW = 2\times F_{\max}$$

Where F_{max} is the highest frequency contained in the message signal.

Frequency Modulation

The FM signal is generated by altering the frequency of the carrier signal. The amount of carrier frequency shift is called the *deviation,* and is proportional to the amplitude of the message signal. The FM signal formula is

$$v(t) = \text{Cos}\ [\omega_c + d\ \text{Cos}\ \omega_m t]t$$

where

$$d = \text{peak deviation (rad/s) or}$$

$$d^1 = \frac{d}{2\times\pi} = \text{peak deviation (Hz)}$$

The modulation index is the deviation divided by the message signal frequency and is

$$m = d/\omega_m \text{ or } d^1/F_{\text{m}}$$

Narrowband FM (NBFM) is where $m < \pi/2$. Wideband FM (WBFM) is where $m > \pi/2$. The time and frequency domain representations are given in Figures 4.2 and 4.3.

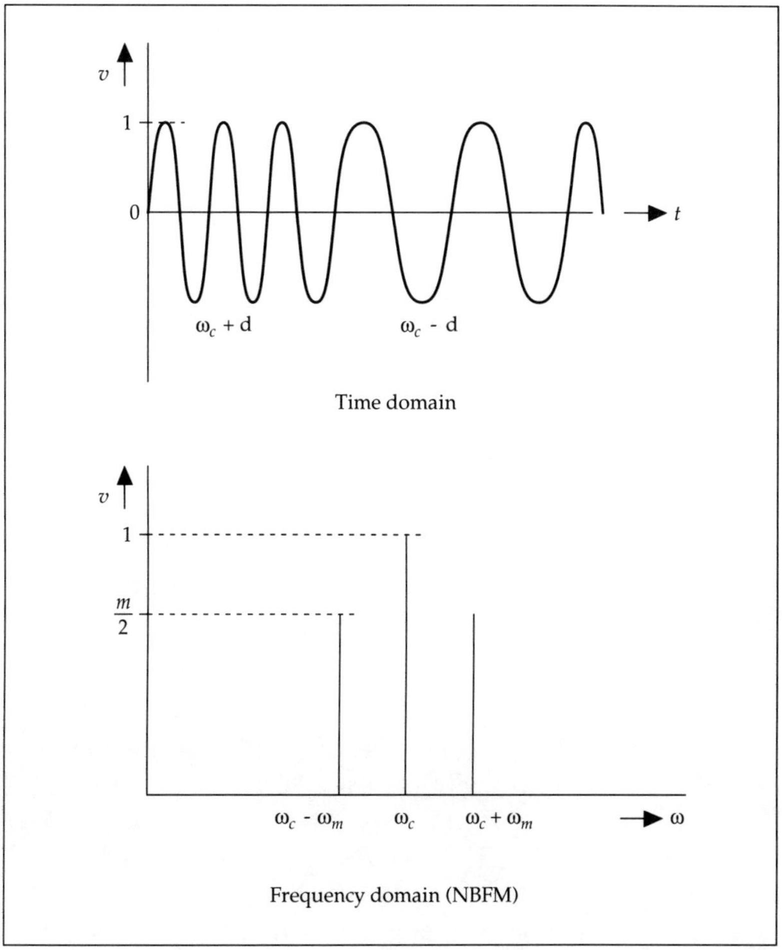

Figure 4.2

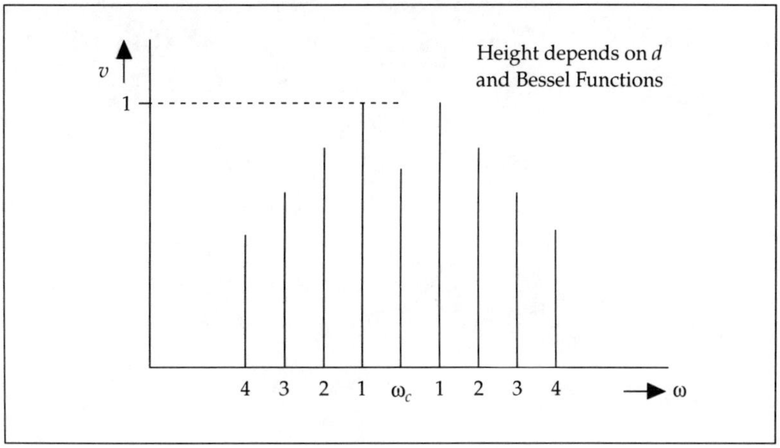

Figure 4.3 *Frequency domain (WBFM)*

With NBFM the bandwidth is:

$$BW = 2 \times F_{max}$$

With WBFM the deviation is included. The bandwidth is:

$$BW = 2 \times (F_{max} + d^1) \text{ [Carson's Rule]}$$

Where F_{max} is the highest frequency in the complex signal and d^1 is the deviation.

Modulation Index

The modulation index is considered further, since its dependency varies according to the modulation method.

With AM or PM (phase modulation), the modulation index is independent of the message signal frequency.

With FM, it is the carrier frequency deviation that is independent of the message signal frequency, and therefore the modulation index varies as a function of that frequency. The relationship is, from above,

$$m = d \,/\, \omega_m \text{ or } d^1 / Fm$$

Now, one popular method used to generate FM is with a phase modulator and an integrator in the message signal path. As it is d^1 that must remain constant with FM, then the modulation index m must increase as the message frequency Fm decreases. As Fm approaches 0 Hz (dc), then m approaches infinity. If implemented by a phase modulator then m is defined as the phase deviation, and it follows that m must be doubled each time that Fm halves. This is a slope of – 6 dB/octave, which a perfect integrator supplies. A phase modulator arranged to give FM is shown in Figure 4.4.

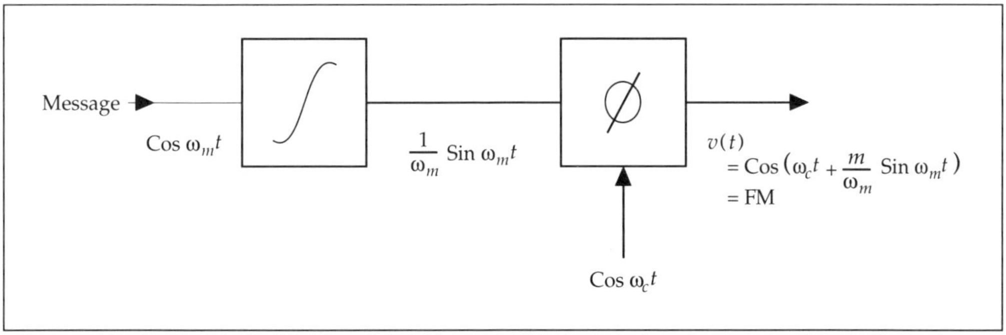

Figure 4.4 *Frequency modulation using a phase modulator*

DIGITAL MODULATION

Binary Amplitude Shift Keying (ASK)

This just means that the amplitude of a continuous signal is shifted in sympathy with 1s and 0s in a binary pattern. If one of the levels is zero, it becomes on-off keying (OOK). OOK is used commonly on fiber optic systems.

With a binary signal, the highest frequency in the message is invoked during a pattern of 01010101... This signal can be 3-state (ternary) or more, not just binary, in which case the pattern giving the highest frequency will be different. The wave in Figure 4.5 represents a binary 101010 pattern. This is the case for a string of 1s and 0s. For random binary data, we have to assume a power spectral density function (PDF) because the signal frequencies do not occupy positions, they are spread out as shown in Figure 4.6.

Now however we wish to represent the data pattern by an ASK signal. For a 101010 sequence it might be depicted as in Figure 4.7. With random data (at the same speed) it will look like Figure 4.8.

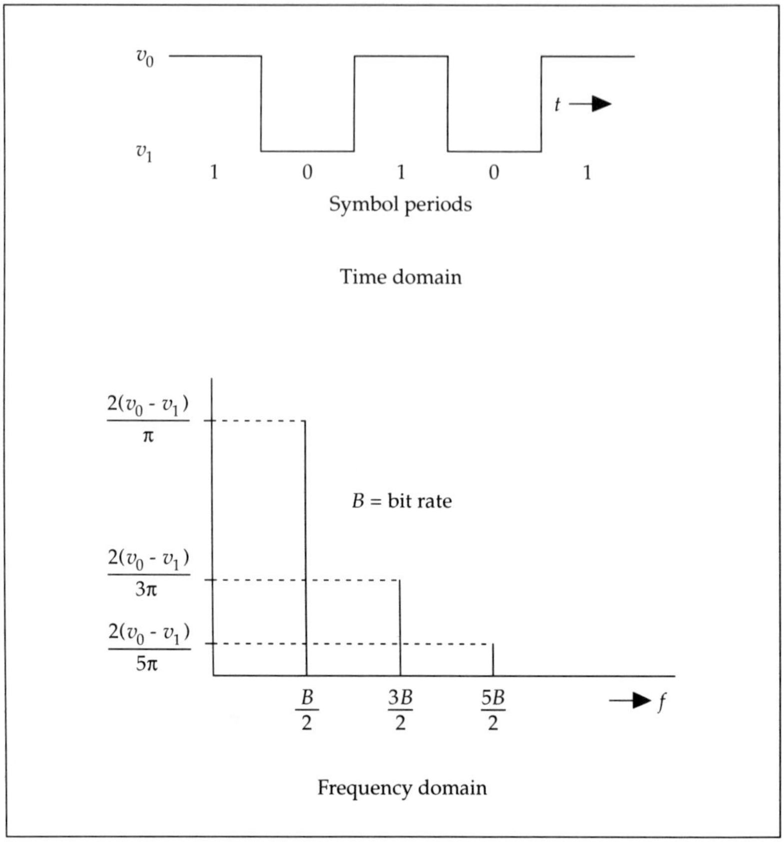

Figure 4.5

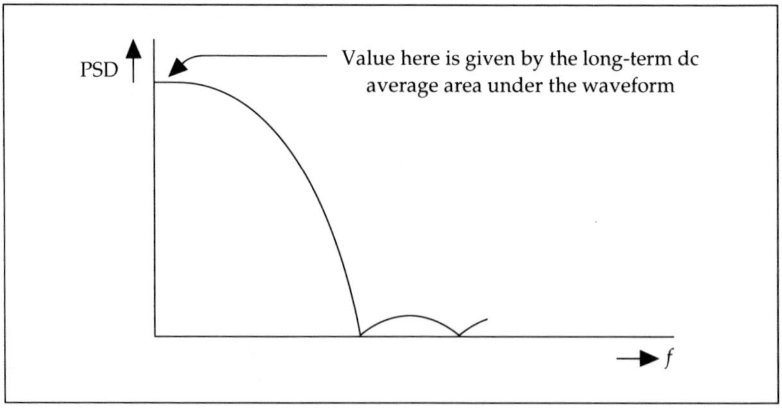

Figure 4.6

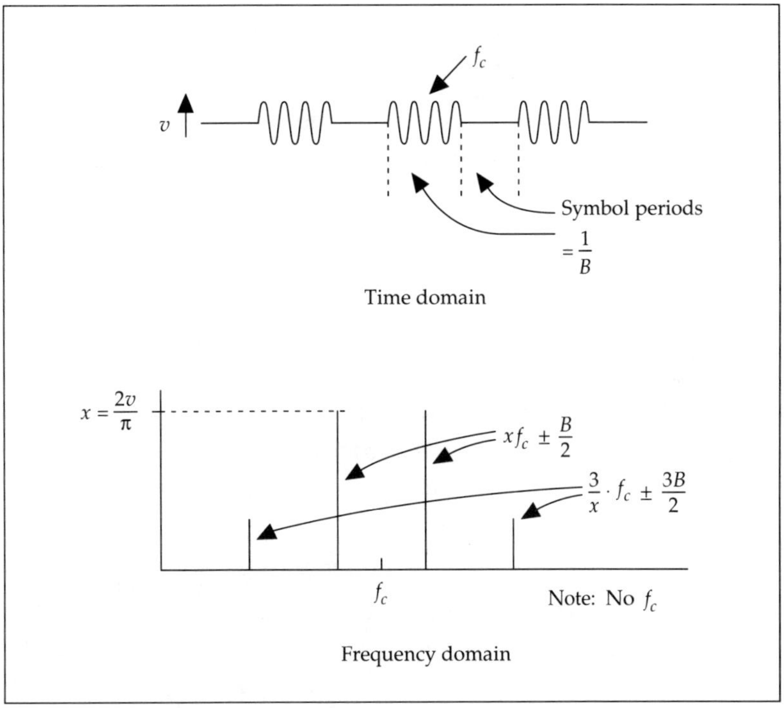

Figure 4.7

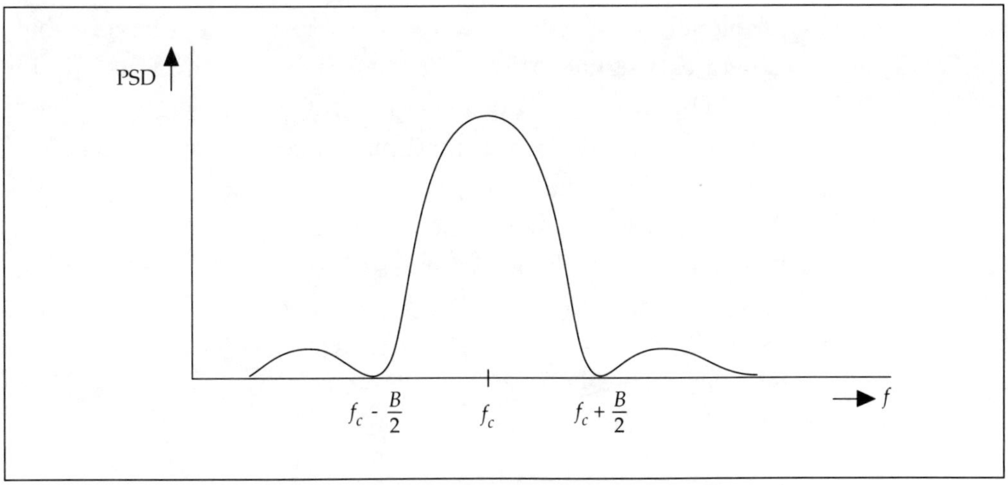

Figure 4.8 *Spectrum of random ASK data*

The bandwidth, then is

$$BW = 2 \times B$$

where B = bit rate.

Phase Shift Keying (PSK)

PSK can easily be derived from ASK by thinking of "filling in" the "no signal" time periods with a signal of 180° shift. The bandwidth taken up by PSK is the same as ASK, that is,

$$BW = 2 \times B$$

where B = bit rate.

NOTE There is no signal at the carrier frequency itself. This is because it spends half the time being multiplied by +1 and the other half being multiplied by −1 and so cancels out. The time and frequency domain signals are as shown in Figure 4.9.

The fact that no clock signal component exists causes problems at the receiver. Generally, it is necessary to extract a clock signal that is synchronous to the data in order to detect the binary states, and if there is no center frequency, there is nothing for the clock regenerator to lock to. One way out of the problem is to employ a frequency doubler at the receiver, which generates a signal at twice the clock frequency, and then divide by 2. This will give a clock signal but in either of two phases with respect to the incoming signal. A unique word is therefore sent to establish data polarity.

Another way out of the problem is to employ differential PSK (DPSK), which only changes state when the last bit and the current bit are different. This is generated by a circuit such as is shown in Figure 4.10. The bandwidth of DPSK is also equal to twice the bit rate.

Frequency Shift Keying (FSK)

FSK is where discrete tone frequencies are allocated to each digital state. In the case of binary, two frequencies are allocated. Multi-FSK (called M-ary) signals can be generated where M is commonly up to 32. If 2-FSK is considered, and the two tone frequencies are called F_1 and F_2, then the distance between them cannot be less than half the bit-rate B, and it is usually 0.7.

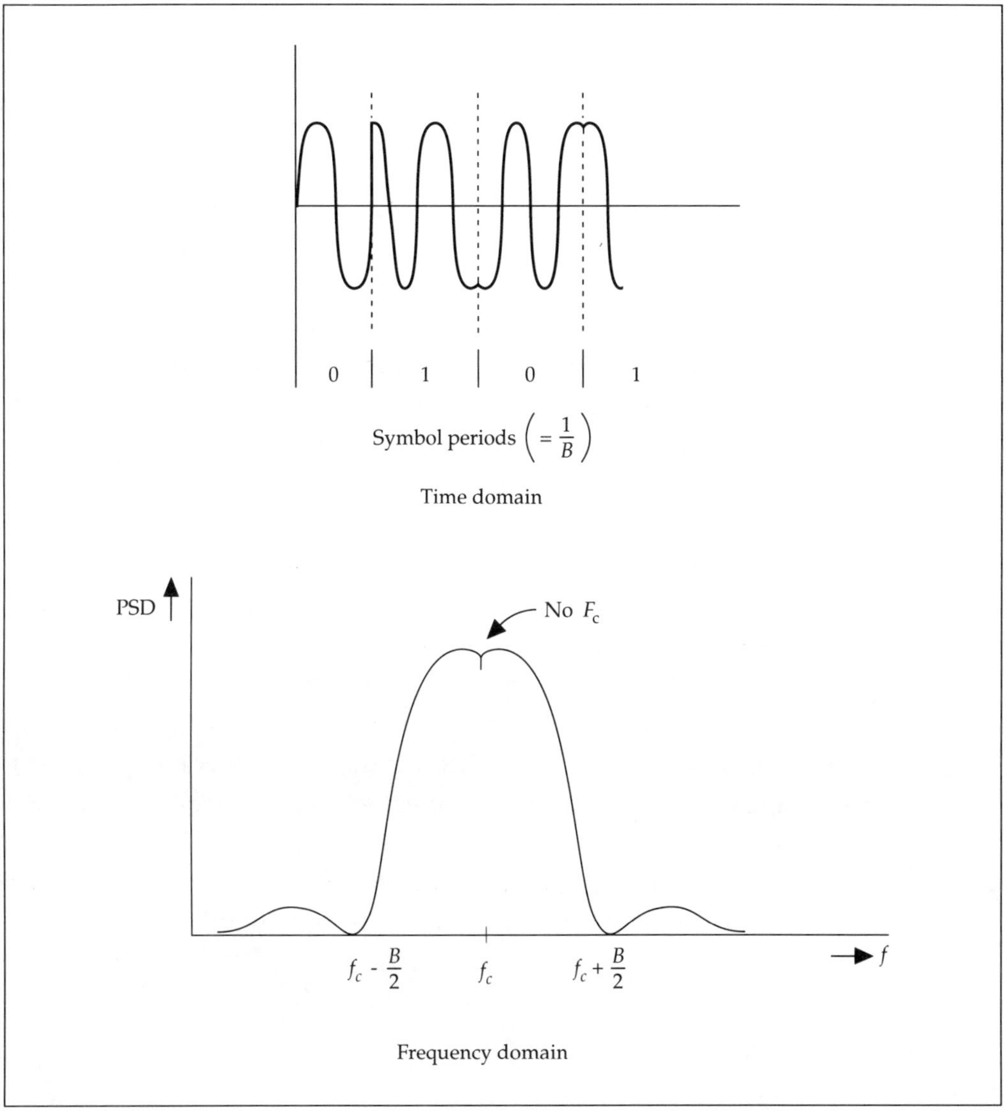

Figure 4.9

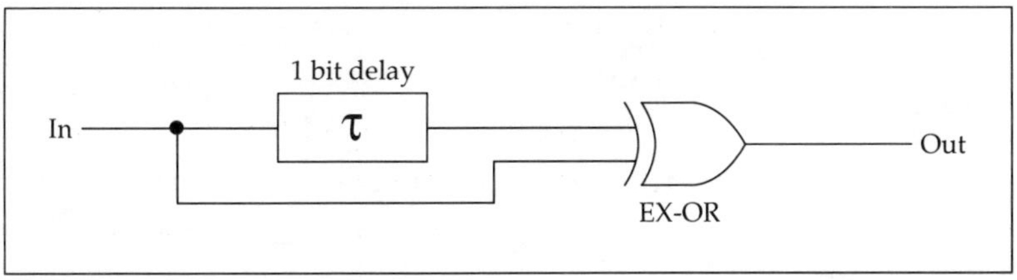

Figure 4.10

An example of the tightest arrangement is fast FSK (FFSK) where:

$$F^1 = 1200 \text{ Hz and represents } 1$$

$$F^2 = 1800 \text{ Hz and represents } 0$$

$$B = 1200 \text{ bits/s}$$

The distance between F_1 and F_2 is 600 Hz, which is equal to $B/2$, and this is the minimum distance allowed. For this reason, FFSK is so-called because it signals as fast as it can for the distance used. As it also uses the minimum distance, and hence minimum spectrum width, it is also sometimes called minimum FSK (MFSK).

FFSK and MFSK are also well-suited to radio use because the state transitions occur while the signal is passing through zero, and hence it is a phase-continuous signal.

FFSK is not the ideal arrangement for immunity to intersymbol interference. An improvement in ISI can be gained by spacing F_1 and F_2 by 0.7 times the bit rate. For 1200 bps this means that the distance is ideally equal to or greater than 840 Hz.

The time and frequency domain representations of the signal are indicated in Figure 4.11. The bandwidth is

$$BW = F_2 - F_1 + 2B$$

Modulation

In the signals considered so far, the "carrier" can either be an rf or optical signal whose frequency is suitable for transmission along cables, fibers, or via aerials and the ether. It can also be an audio (baseband) frequency chosen such that the bandwidth fits within the audio band for transmission using a system primarily designed for speech. In these cases it must be modulated on to an rf carrier for transmission over radio. Remodulation

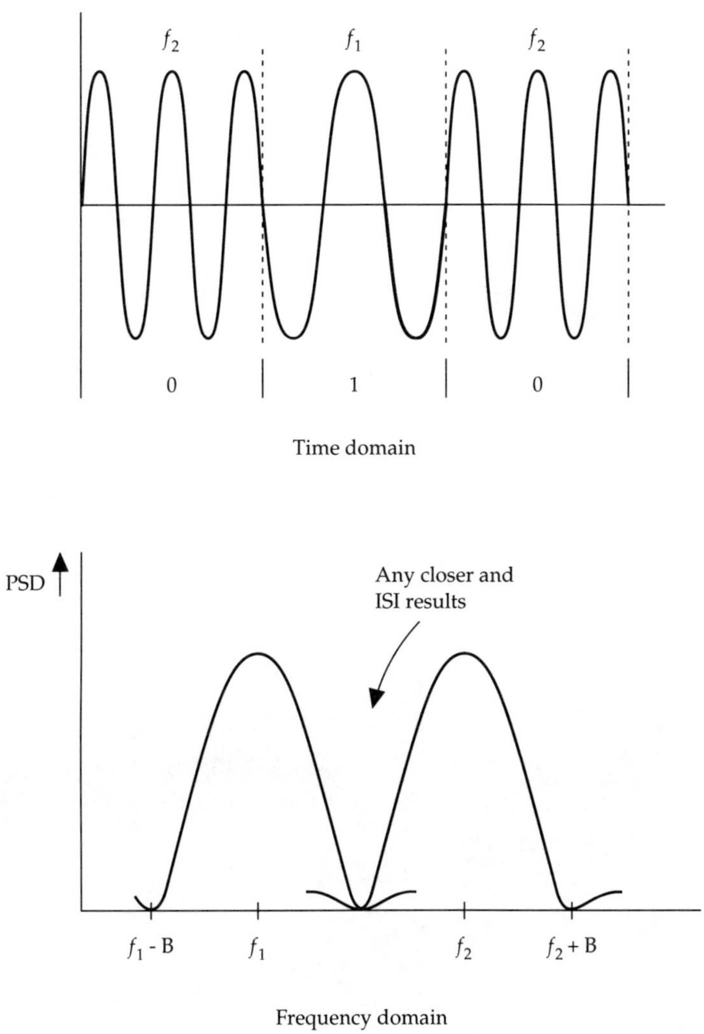

Figure 4.11

onto an rf carrier also makes the system more resistant to interference.

Any of the above systems, ASK, PSK or FSK, can be used with a "carrier" at audio frequencies and modulated onto an rf carrier signal using AM or FM.

Table 4.1

	ASK, PSK	**FSK**
Baseband	$2B$	$F_2 - F_1 + 2B$
AM	$2(F_c + B)$	$2(F_2 + B)$
NBFM	$2(F_c + B)$	$2(F_2 + B)$
WBFM	$2(F_c + B + d^1)$	$2(F_2 + B + d^1)$

The bandwidths for the various schemes are given in Table 4.1. The frequency domain representations of ASK (or PSK) and of FSK with NBFM (or AM) are given in Figure 4.12.

Examples

1. A 12.5 kHz PMR (private mobile radio) channel carrying speech.

$$\text{Deviation} = 2.5 \text{ kHz max}$$
$$\text{Audio band} = 300 \text{ Hz to } 3.4 \text{ kHz}$$

Both of these limits are given in DTI (Department of Trade and Industry) specification MPT (Ministry of Post and Telecommunication) 1326. The modulation index at the maximum audio frequency is

$$m = d^1/F_{\max} = 2.5/3/4 \text{ which is} < \pi/2$$

So, it is NBFM. The bandwidth is thus

$$BW = 2 \times F_{\max} = 6.8 \text{ kHz}$$

2. A 12.5 kHz PMR channel carrying FFSK

$$\text{Deviation} = 60\% \text{ of system deviation} = 1.5\text{kHz}$$
$$F_1 = 1200 \text{ Hz}$$
$$F_2 = 1800 \text{ Hz}$$
$$B = 1200 \text{ bits/s}$$

These are defined in DTI code of practice MPT 1317.

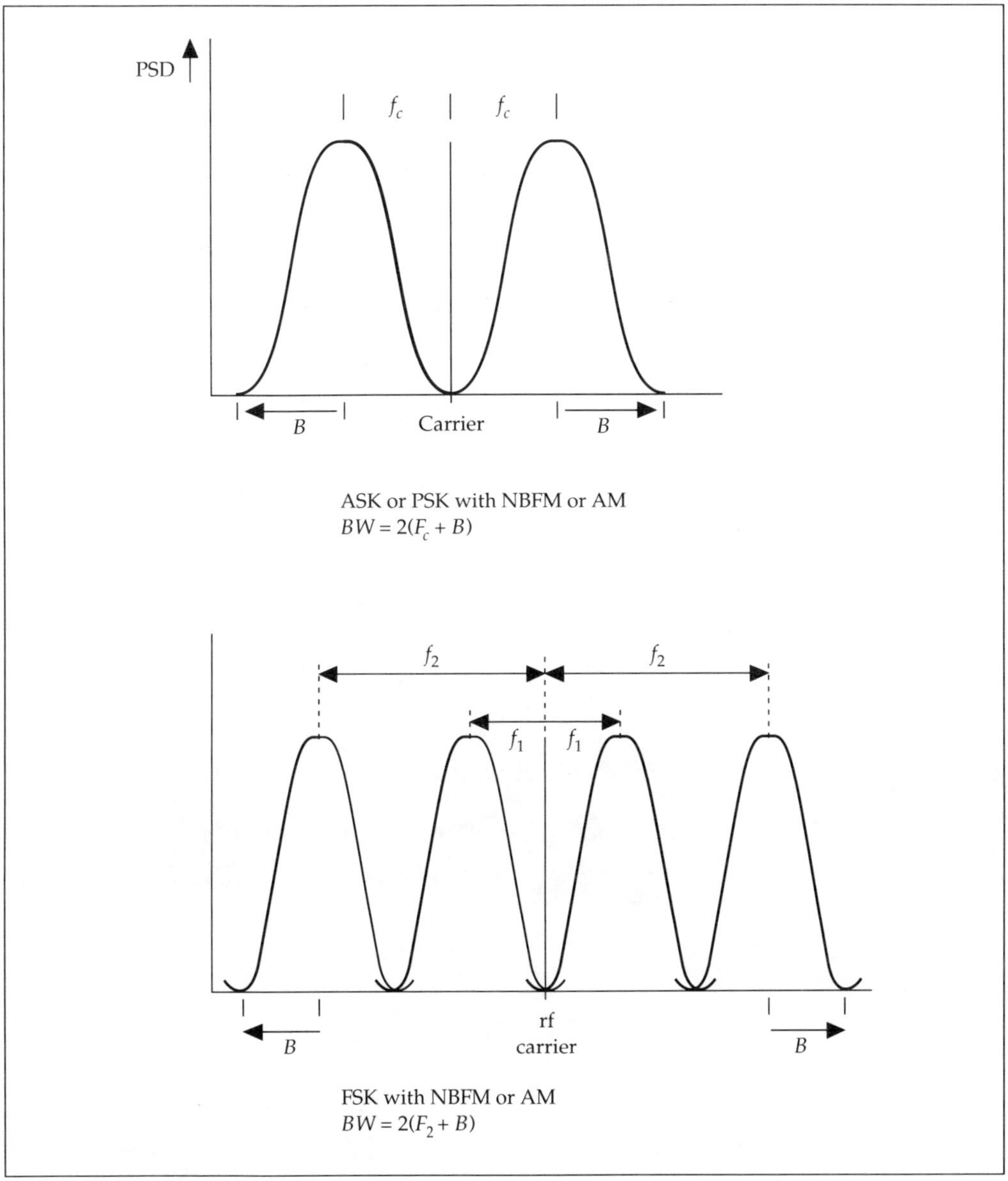

Figure 4.12

$$F_{max} = F_2 + B = 3 \text{ kHz}$$

The modulation index $m = d^1/F_{max} = 1.5/3.0$ which is $< \pi/2$. So we have NBFM. The bandwidth is thus

$$BW = 2(F_2 + B) = 6.0 \text{ kHz}.$$

In 12.5 kHz channel PMR receivers, the IF bandwidth is typically 7.5 kHz, which accommodates either speech or FFSK signaling with some margin to avoid excessive phase distortion brought about by the filter edges. The margin also allows for inaccuracy of rf oscillator frequencies at both ends.

3. A cellular channel carrying speech

$$\text{Deviation} = 9 \text{ kHz}$$
$$\text{Audio band} = 300 \text{ Hz to } 3.4 \text{ kHz}$$

In this case $m = 9/3.4$ which is $> \pi/2$. So, it is WBFM. The bandwidth is

$$\text{BW} = 2(F_{\text{max}} + d^1) = 2(3.4 + 9) = 24.8 \text{ kHz}.$$

So the cellular networks use up their entire 25 kHz allocation, making adjacent channel working impossible from the same or adjacent cells.

4. A cellular channel carrying FFSK

$$\text{Deviation} = 60\% \text{ of system deviation} = 5.4 \text{ kHz}$$
$$F_1 = 1200 \text{ Hz}$$
$$F_2 = 1800 \text{ Hz}$$
$$B = 1200 \text{ bits/s}$$
$$F_{\text{max}} = 1800 \text{ Hz} + 1200 \text{ Hz} = 3 \text{ kHz}$$

Modulation index $m = d^1/F_{\text{max}} = 5.4/3 = 1.8$ which is $> \pi/2$. So we have WBFM. Bandwidth is $2(F_2 + B + d^1) = 2(1800 + 1200 + 5400) = 16.8$ kHz.

5. A Band III channel carrying FFSK

1200 Hz	
1800 Hz	
Subcarrier	1500 Hz
Bit rate	1200 bit/s

Occupied bandwidth 600 Hz
Centered on 1500 Hz; i.e., 900 to 2100 Hz
Modulated on to 12.5 kHz
1.5 kHz deviation.
Narrow band – deviation 1.5 kHz
Max frequency of information 2.1 kHz.

$$\begin{aligned}\text{i.e., bandwidth} &= 2[F \text{ tone} + (0.5 \times \text{bit rate})]\\ &= 2[1500 + 0.5 \times 1200)]\\ &= 4.2 \text{ kHz}\end{aligned}$$

6. Using modems

Bandwidth required about subcarrier tone.
1500 (FFSK) or 1950 Hz
Remodulated onto rf carrier

$$\text{Bandwidth} = 2(F \text{ tone} + (0.5 \times \text{bit rate}) + F_d \text{ for WBFM}$$

F tone = Subcarrier tone frequency

F_d = Max deviation of carrier

(for NBFM delete F_d).

7. DFSK for POCSAG (Post Office Standardization Advisory Group)

Bandwidth = 0.5 × bit rate.

Example

Define bandwidth of 2.4 kb/s signal.

1. Put onto carrier using DFSK
2. Processed by a modem using a subcarrier of 2.2 kHz before being modulated onto an rf carrier at max deviation of 1.2 kHz.

1. Bandwidth = 0.5 × bit rate
   = 1.2 kHz
2. Bandwidth = 2 [2.2 + (0.5 × 2.4) + 3] kHz
   = 12.8 kHz.

i.e., a modem increases bandwidth, but increases immunity to interference and rf parameter drift.

8. Digital modulation

QPSK

Total bandwidth = 2 × 0.25 × bit rate
= 0.5 × bit rate

Spectral efficiency = 2 bits/Hz

Same bandwidth for FFSK, DFSK, GMSK.

(DFSK puts modulation straight onto carrier).

CHAPTER 5

CIRCUIT TECHNOLOGY

RECEIVERS

The receiver configuration that has been used for professional communications for the last 30 to 40 years is the *superhetrodyne* or *superhet*. The name arises from the circuit technique used: "super" being an abbreviation of supersonic, an indication that a method is used that is above audio and "hetrodyne" a mixing or beating together of two signals. The typical receiver system for speech is shown in Figure 5.1. The superhet principle is applicable to both AM and FM.

We will outline the main features of the superhet and then develop an analysis of the problems of this approach (particularly with respect to data systems).

The superhet technique converts the incoming carrier down to a lower frequency (keeping the original modulation

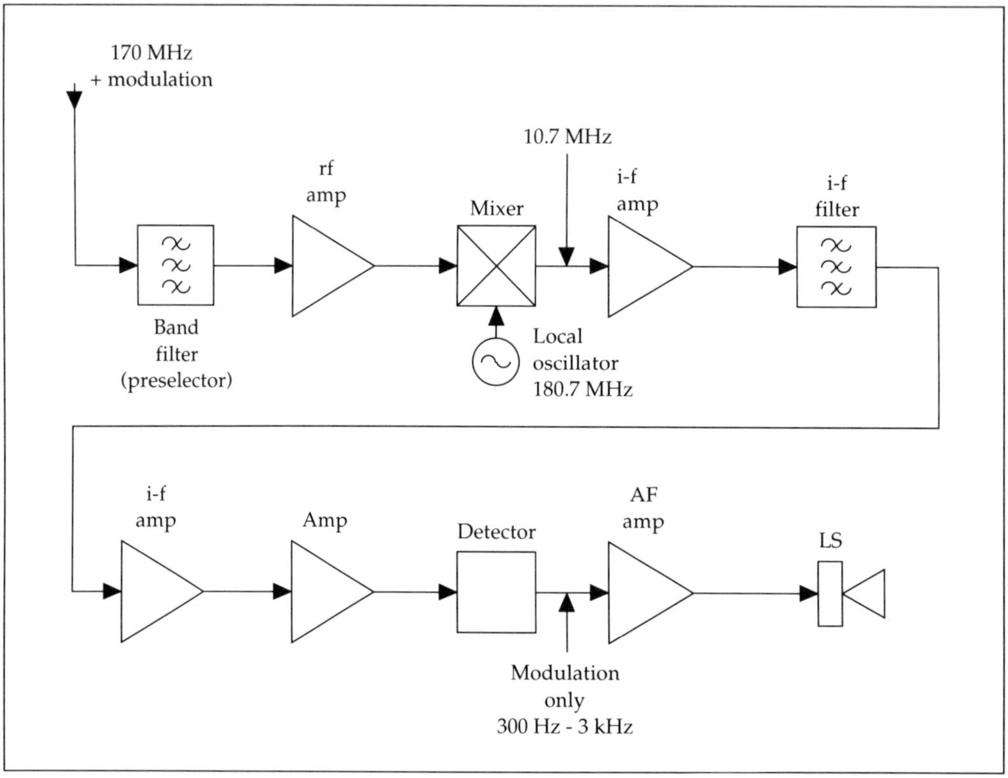

Figure 5.1

unchanged) at an early stage in the receiver. This lower frequency is called the *intermediate frequency* (i-f) because it is between the carrier and final audio frequencies. It is at the i-f that most of the signal processing occurs—amplification, selective filtering, leveling, and detection. This is for two main reasons:

1. To overcome certain difficulties that occur in trying to process the signal at its original transmited frequency.
2. To enable the maximum use of low-cost, low-power consumption circuits (both integrated and discrete).

A key circuit element or component in the superhet is the mixer. An examination of its workings will highlight some of the disadvantages and problems that have to be overcome in its use.

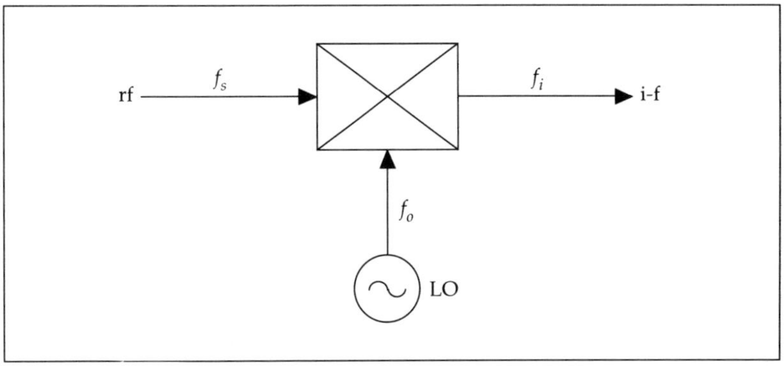

Figure 5.2 *The mixer*

THE MIXER

The object of the mixer is to mix, or beat, two input signals to produce an output that is a function of the input frequencies.

In a superhet receiver (See Figure 5.2), f_s is the original transmited signal as intercepted by the receiver antennae with some filtering and amplification. f_o is the signal generated by an oscillator within the receiver itself (hence referred to as the local *oscillator* [LO]).

If we first consider the input as a single frequency, i.e., just one channel, we can examine the mixer action.

In order to achieve a mixing or frequency conversion process the mixer must be a nonlinear device. The desired output signal at f_i is:

$$f_s + f_o \text{ or } f_s - f_o$$

The ideal mixer should fulfill certain criteria:

1. Have a transfer function that contains no terms higher than the second order (i.e., exhibits a square-law characteristic). This is to prevent the mixer itself generating frequencies other than the sum and difference.
2. Have a conversion gain rather than a conversion loss—to improve the noise performance of the receiver.

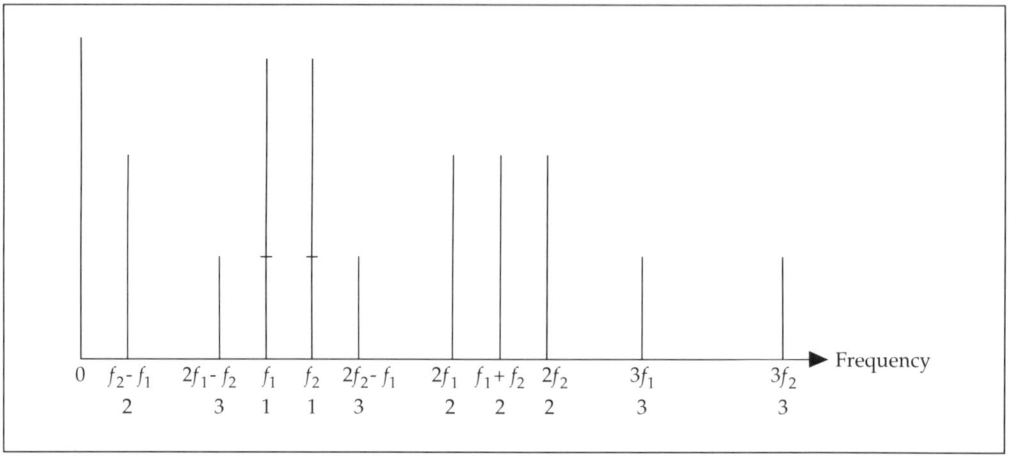

Figure 5.3

3. Have a high degree of isolation between any two ports of the device—to prevent unwanted interactions.

As mixers in practice do not have a perfect square-law characteristic intermodulation products are output as well as the desired frequency.

Consider two signals reaching the mixer via the receiver front end f_1 and f_2 (both labeled as 1 in Figure 5.3). They are shown as having the same amplitude although in practice this is unlikely to be the case. If one or both signals are sufficient in amplitude to overload the mixer, then other unwanted signals will be produced. These unwanted signals are shown and labeled as type 2 or type 3 signals. These signals are the second-order and third-order products, respectively. It can be seen that the third-order products produced lie very close to the wanted signal area and will therefore be translated down into the i-f passband. If they are too close to the wanted channel in the i-f band then the i-f filter may have insufficient selectivity to reject them.

The other point to keep in mind is that in practice most communication takes place in multichannel bands. This means that, as well as the wanted channel frequency being converted into i-f passband, all other channels are similarly converted and

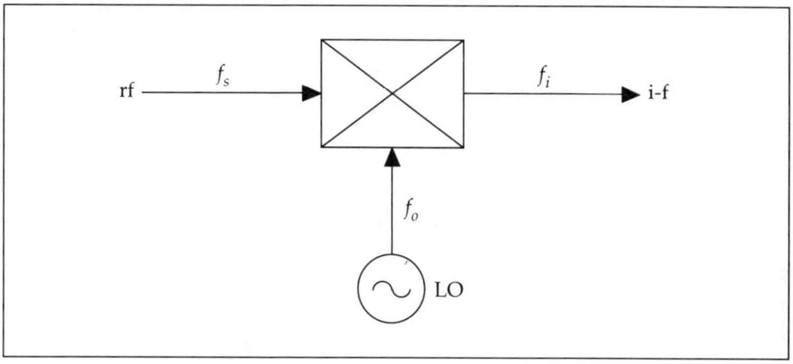

Figure 5.4 *The mixer*

hence unwanted signals may lie 25 kHz, 12.5 kHz, or even 6.25 kHz away from the i-f. This again requires a very high degree of i-f selectivity to keep them from reaching the detector.

To return to the function of the ideal mixer, the output will consist of the sum and difference frequencies of the inputs.

In Figure 5.4,

$$\text{rf input (wanted)} = f_s$$
$$\text{LO freq} = f_o$$
$$\text{mixer output} = f_i$$
$$f_i = f_o + f_s$$

e.g.,

$$f_s = 170 \text{ MHz}, f_o = 180 \text{ MHz}$$
$$f_i = f_o + f_s = 350 \text{ MHz}$$
$$f_i = f_o - f_s = 10 \text{ MHz}$$

There is, however, a second input frequency that will be converted to the i-f. If the receiver i-f (fixed by the i-f filter) is 10.7 MHz and the LO is 140.0 MHz then the mixer will convert both 140.0 MHz + 10.7 MHz = 150.7 MHz and 140.0 MHz – 10.7 MHz = 129.3 MHz into the i-f passband. The receiver has no way of comprehending which of these two is the wanted frequency. This unwanted response is called the *image frequency*. The only way to prevent this image frequency entering the i-f and hence the detector is to prevent the unwanted frequency from reaching the mixer. This is a prime function of the front-end filtering of the receiver. The *preselect filter* (and any other filtering circuits before the mixer) must have sufficient attenuation at the image frequency to enable the receiver to meet its specification. A typical receiver image rejection specification might be 75 dB (image relative to wanted signal).

Examination of the figures will show that 150.7 MHz – 129.3 MHz = i-f × 2. It follows therefore that the higher the i-f the further away is the image frequency and hence the simpler the front-end filtering. However, we have seen that we require a very narrow bandpass characteristic from the i-f filter to maintain adjacent channel performance and remove any other unwanted signals. This becomes more demanding as the i-f rises—the design engineer is caught between the two constraints.

The classical superhet approach to the problem is to divide the difficulty across more circuit elements. A technique referred to as a *double superhet* configuration is used. (See Figure 5.5.)

In this technique, the first mixer produces a relatively high i-f (21.4 MHz is typical). This means that the image will be 21.4 × 2 = 42.8 MHz away from the channel signal. The design and number of the preselect filter stages can be reduced, hence the component count and setting up time is kept to a minimum. Further down the chain a second mixer is used to produce a second i-f. The filter between the two mixers can be simple in design, since the wanted and image frequencies are fixed. Due to the "soft," all-enveloping shape of this filter, it is often referred to as a *roofing filter*. The second i-f will be as low as possible to ease the main selective filter design—455 kHz is a common choice. This can then be a ceramic or tuned filter rather than a more expensive crystal filter. Similarly the FM discriminator is rather easier to implement. There are of course, problems:

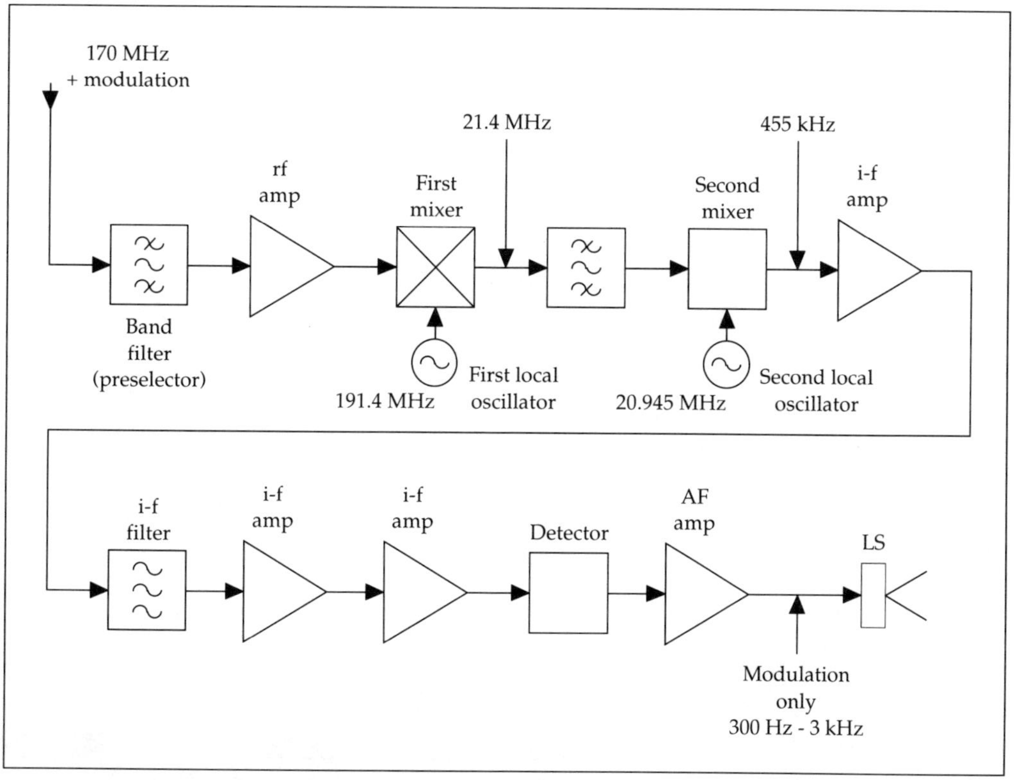

Figure 5.5

1. The receiver is more complex, increasing current consumption and decreasing reliability.
2. The receiver with more components is physically larger.
3. There are now two oscillators within the receiver resulting in more problems of drift and instability.
4. The increased number of frequencies within the receiver have much more potential for intermodulation products to be produced—both in-band and out-of-band.
5. Due to the large number of stages between the antenna and the high selectivity filter, it is very difficult to attenuate large, close-in signals. This means all stages up to this point must have a wide dynamic range (i.e., good linearity).

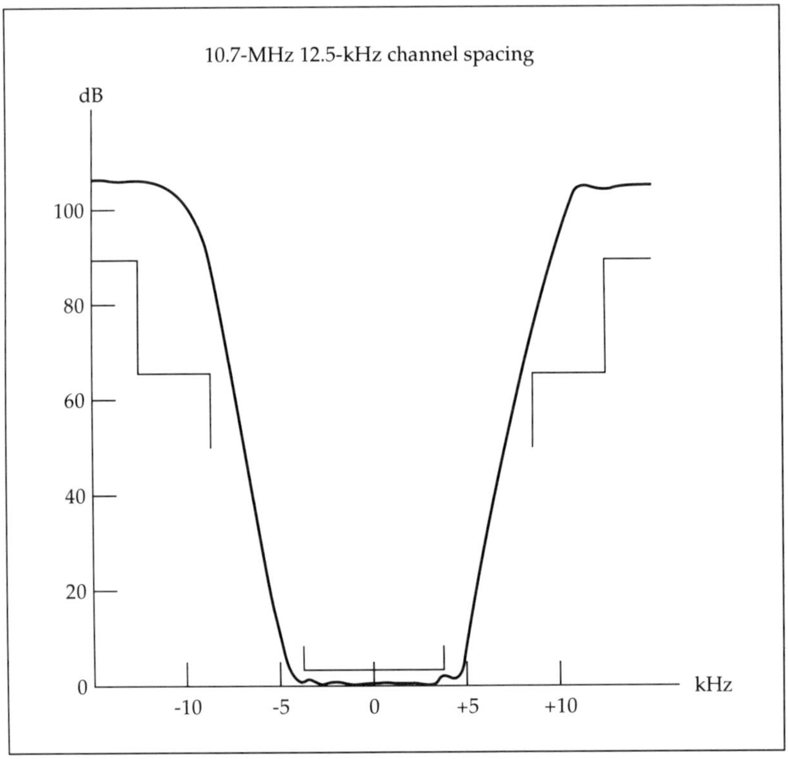

Figure 5.6

THE i-f FILTER

The i-f filter provides the very high degree of selectivity required to enable the superhet receiver to meet its spurious signal rejection specifications. These can be signals from adjacent channels, received interfering signals, or intermodulation products generated in the receiver.

The ideal filter will have a flat minimum attenuation response in the i-f passband with steep sides rising rapidly to the stop band frequencies.

The ratio of response at low levels of attenuation (usually 6 dB) to high levels of attenuation (usually 60 dB) by frequency is referred to as the *shape factor*. It is important that the stop attenuation continues out above and below the i-f to at least the detector bandwidth. If the stop attenuation falls at any point, then signals at this frequency will break through to the detector. (See Figure 5.6.)

The difficulty comes in the superhet in trying to make a filter meet this specification. Due to the high ratio of center frequency to bandwidth (10.7 MHz i-f to 9 kHz bandwidth is not unusual), it is not possible to realize this filter shape using conventional inductor/capacitor techniques. A frequently used filter to meet the requirement in the superhet application is the ceramic filter. This filter relies on a phenomenon known as the *piezoelectric effect*. Its molecular lattice distorts under electrical stress. This effect can be utilized to create a filter with the required performance. This performance will however only be realized if the filter is used in precisely the correct manner. The matching of the input and output determine the correctness of the filter response. Incorrect matching of the filter will cause phase linearity problems, which manifest themselves as variable time delays through the filter with changes in frequency. These time delays or phase distortions can cause severe timing errors at the transition states of digital waveforms. The "skew" will in turn cause bit errors to occur in data passed through such a system. This characterization of filters according to their phase response is referred to as

$$\text{the group delay time (GDT)} = \frac{d\,\theta}{d}$$

(See Figure 5.7.)

Basic filter types are shown in Figure 5.8 with pass response and GDT.

Review

We have outlined the basic concept of the present commercial receiver configuration—the superhet. We examined some details of the mixer, showing its design aims and functions, and discussed the problems of image frequency related to the choice of i-f and also spurious frequencies generated from within the mixer. We considered another important component—the filter—and highlighted its exacting requirements. And finally, we

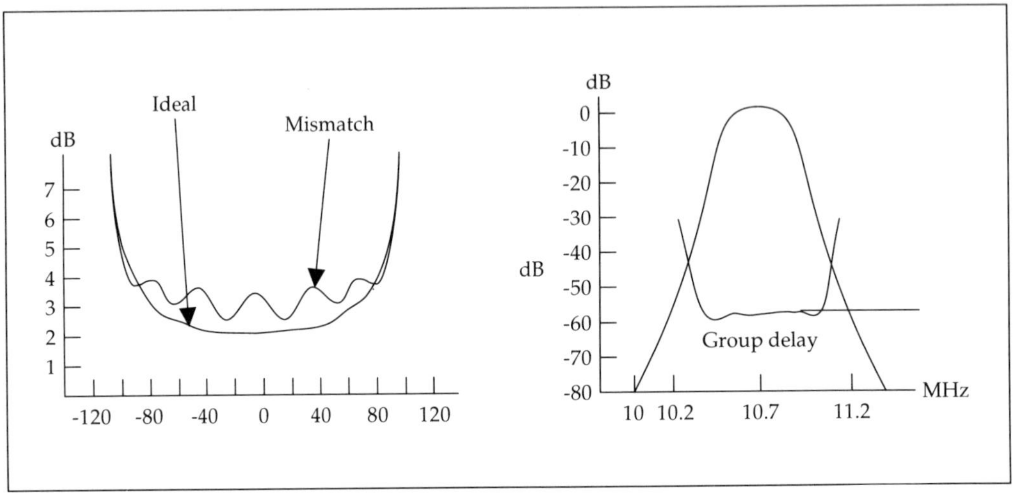

Figure 5.7

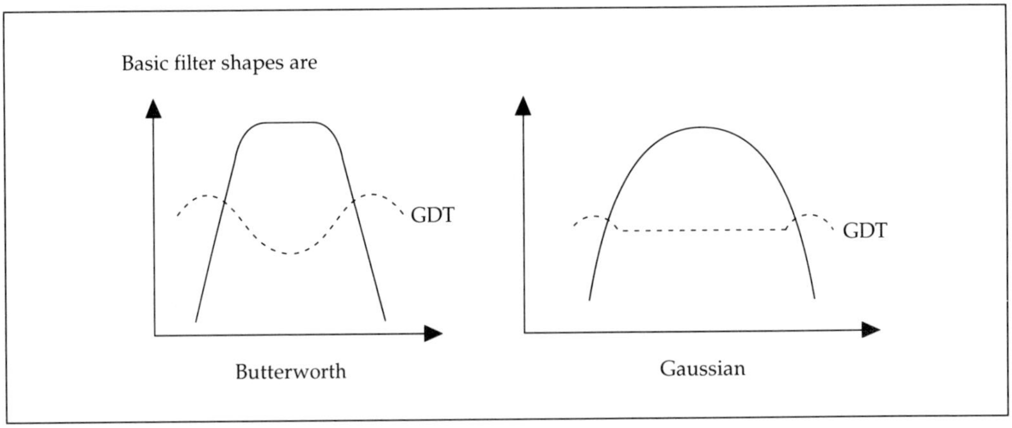

Figure 5.8

proposed the double superhet as a solution but showed it to be a compromise.

An Alternative Approach

An approach used to overcome the image problem (particularly in HF—high-frequency—receivers), is to make the first i-f higher than the total operating frequency of the receiver. In this way the image frequency falls totally out of the operating range of the system and hence the preselect filtering is easily

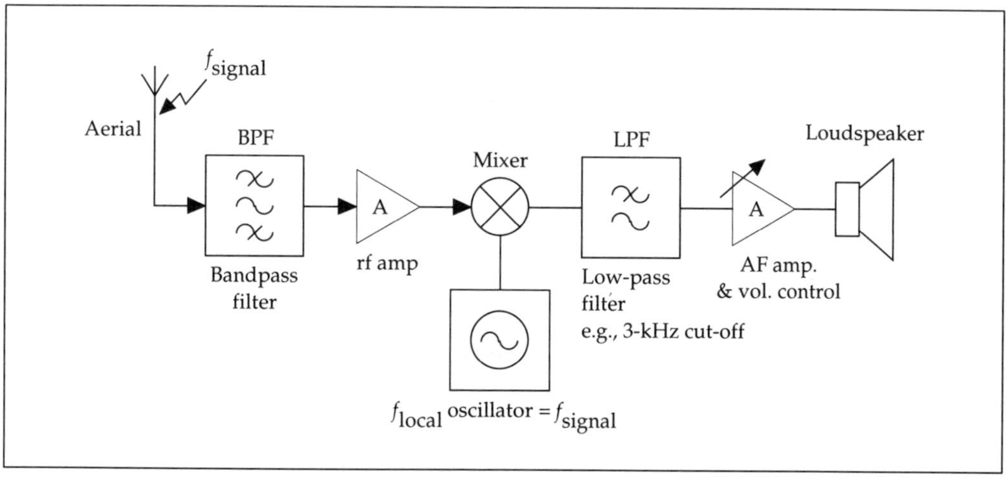

Figure 5.9

achieved. This technique cannot easily be applied to VHF as the i-fs become high.

Another approach that was proposed in the 1940s was to use a superhet configuration but with the local oscillator running at the received signal frequency—the so-called *direct conversion receiver*. The output from the mixer would be subject to low-pass filtering to obtain the audio frequencies. These audio frequencies will be the original modulating signal. (See Figure 5.9.)

The receiver, in its basic form uses a relatively simple preselect filter to stop large out-of-band signals that would otherwise cause intermodulation problems from reaching the i-f amplifier and mixer. Image rejection performance is not necessary, since the i-f equals 0 Hz, the image and wanted frequencies are coincidental, and so a rejection of a separate image frequency is not required.

This highlights the main problem with this simple approach: Modulation systems that utilize double sideband methods—AM, FM, dsb, etc.—will not work with this simple system. Since the difference between wanted signal and the image is zero, a signal 500 Hz above the carrier will be received as will a signal 500 Hz below the carrier. In other words, the direct conversion receiver (or zero i-f receiver) cannot distinguish between upper and lower sidebands. To express the problem mathematically, the received signal may be represented as a voltage $e(t)$ where

$$e(t) = y(t)\cos(\omega t + \theta)$$

$y(t)$ is the modulating function

ω is the carrier frequency

θ is the arbitrary phase angle

the local oscillator generates a voltage $e'(t)$ where

$$e'(t) = a\cos\omega t$$

a is a constant

After mixing and filtering a signal is obtained $e''(t)$ where

$$e''(t) = a'\,y(t)\cos\theta.$$

The term cos θ causes a problem. This is a phase component that is generated outside the transmit/receive system. As θ depends on the distance between the transmitter and receiver and also on the propagation conditions, cos θ can take any value between 0 and 1.

This term will remove all validity from the modulation. An approach suggested to overcome this problem uses a "relative" rather than an "absolute" method of conversion.

In this system two identical receiving channels are constructed using local oscillator injection having a quadrature (90°) phase difference. The outputs from the two mixer channels will still however contain the θ term. Because of the quadrature nature of the oscillator injection, this term will also be at 90° phase difference in both channels; i.e., Sin θ and Cos θ.

So the complete converted signals, $e_1(t)$ and $e_2(t)$, become

$$\text{where } e_1''(t) = a'\,y(t)\cos\theta$$

$$\text{and } e_2''(t) = a'\,y(t)\sin\theta$$

See Figure 5.10.

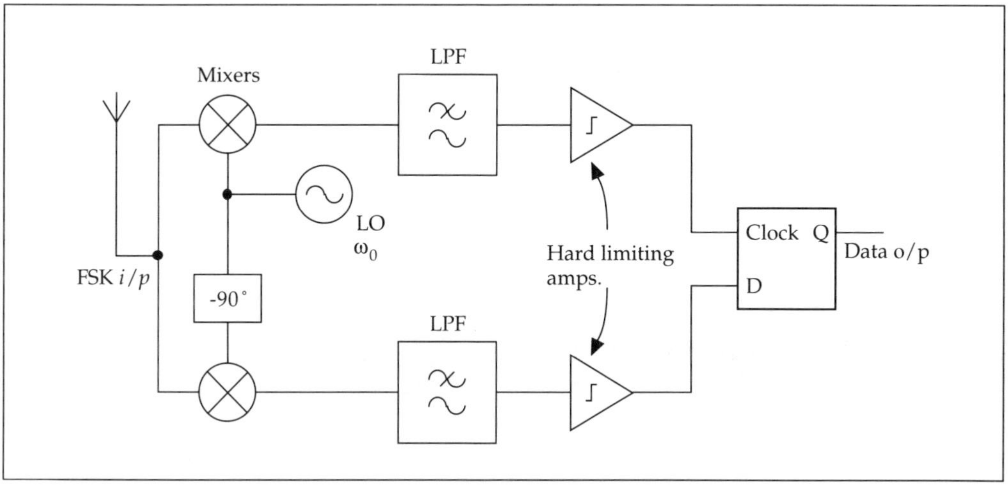

Figure 5.10

There are now two methods of obtaining $y(t)$ independent of θ:

1. Combining the two signals in a network to remove the dependence on θ

OR

2. To utilize the relative magnitude of e_1'' and e_2'' to reduce θ to a standard value, zero, in which case e_1'' is the desired output.

Figure 5.11 shows a practical approach to obtaining the demodulated signal. The method relies on differentiating the Cos A and Sin B terms to obtain Sin D and Cos C respectively. Multiplying Sin by Sin and Cos by Cos gives a waveform at twice the frequency (baseband) with a dc offset equal to half the peak-to-peak value. In the case of the Cos^2 path (upper) the offset is positive and for the Sin^2 path (lower) the offset is always negative. If the lower path signal is subtracted from the upper, the twice frequency term is cancelled and only the dc term (recovered modulation) remains.

Several proposals to achieve the above have been suggested by Barber, Costas, Weaver, and others. These proposals

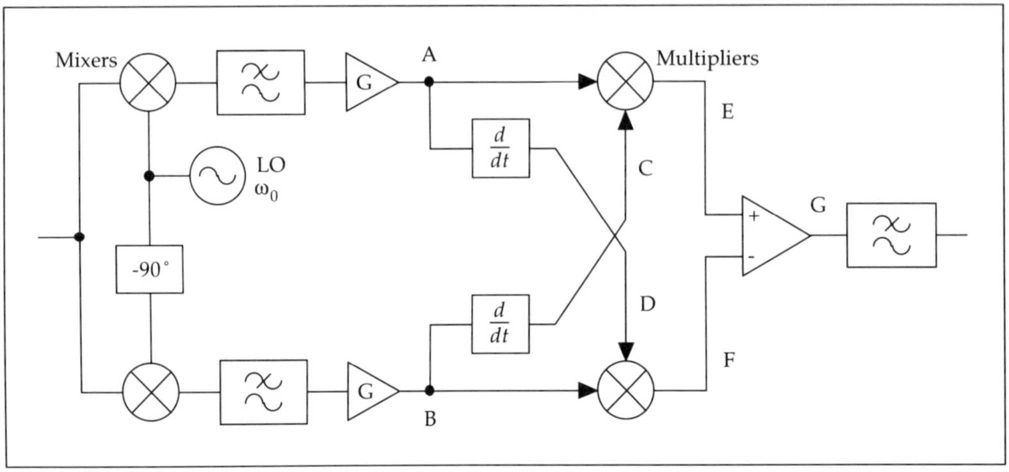

Figure 5.11

were made in the later 1940s. The various modulation recovery processes each have their advantages and disadvantages. The practical problems themselves arise mainly from the differences between the ideal specification for each circuit element and the achievable specification. Problems such as carrier leakage through and around the mixers give rise to dc offsets in the system, thus distorting the recovered information. If the receiver is to be used over any realistic frequency range the 90° phase-shift network must maintain its phase stability over the operational range (also with temperature). Gain inequalities in the two channels can produce problems.

One compromise approach is to use a "near-zero" i-f design as shown in Figure 5.12. The single low i-f overcomes many problems; e.g., the frequency for the 90° phase-shift network is constant, the mixers have better linearity and lower leakage.

However, the image problem returns, so care must be taken in the choice of the i-f and, of course, a preselect filter will again be necessary.

Much of the signal processing requirement is fulfilled by arithmetic procedures; i.e., differentiation, multiplication, subtraction, etc. If the speed or frequency at which this is done can

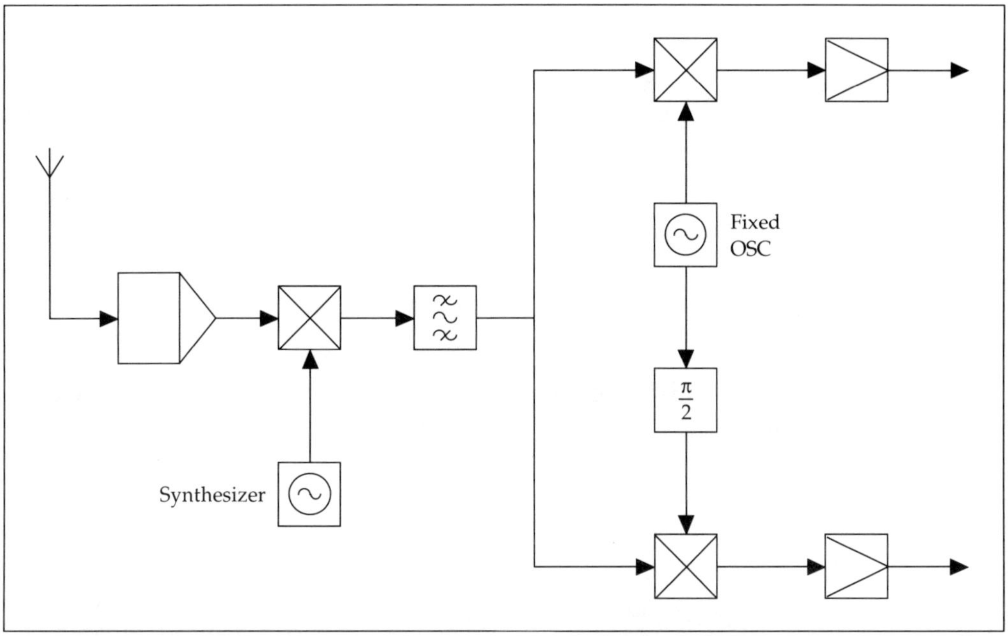

Figure 5.12 *Single i-f receiver*

be brought down to be within the capability of digital signal processors then we have a powerful solution. The frequency of process is, of course, the i-f. However, we saw that low i-fs make excessive demands on the preselect filter. Modern technology provides a solution. The Surface acoustic wave filter or SAW filter. This is a filter/material technology capable of meeting this exacting requirement.

The other thought to be kept in mind is the dynamic range requirements of a commercial receiver. The minimum-usable sensitivity to large signal-handling capability range of the receiver in practical applications is in the order of 100 dB. In the superhet this is taken care of, in the case of FM, by the gain-limiting range of the i-f amplifier. For the AM receiver the AGC (automatic gain control) and detector ranges provide the handling capability. The direct conversion receiver operates in a linear mode through the signal recombination circuits so that it is the following elements that must have sufficient dynamic range, typically 18 bits if digital.

Summary

We have discussed a solution to the problems of the superhet radio. This solution, the direct conversion or zero i-f receiver, has been extensively analyzed in the past and shown to be a potentially simple and elegant approach to rf reception. However, the problems that have been indicated in the direct conversion process have kept its use through the 1950s to 1970s confined to the amateur radio area.

1990s state-of-the-art electronics are now making possible the zero i-f receiver and its use in present and proposed communication systems.

TRANSMITTERS

Discussion of the higher level modulation techniques required to achieve power or spectrum efficiency shows the need for modulation systems with high-phase integrity. It is not uncommon in current systems to produce information-bearing carriers with up to 512 discrete phase states. These phase states may well be produced by modulation techniques that also create large amplitude changes in the carrier. Once modulation has been achieved it is necessary to amplify the power level of the carrier to that required for transmission.

Several approaches are being attempted to handle this problem in the power amplification (PA) stages of the transmitter. Semiconductor device technology is moving forward at a considerable pace, and rf amplification devices are being developed and manufactured with considerably increased linearity. This means that circuit configurations can be simple and effective. A second approach being used relies on predistorting the input to the PA. The distortion on the carrier to be amplified is made equal and opposite to the distortion produced by the PA, so the output will be linear—without distortion. A third technique is the use of feedback; i.e., a sample of the output reinjected at the input to modify the amplification process. Other techniques such as the Cartesian loop transmitter may be considered. This is rather similar to the *quadrature phase technique* used in the direct conversion receiver.

DEVELOPMENT IN FREQUENCY SYNTHESIS

Since frequency synthesizer techniques using phase-locked loops are covered exhaustively in many books and magazines,[1] we will not examine their workings here. However, a few comments covering the latest developments may be useful.

Newer integration techniques put the prescaler (single or dual modulus) on board the main LSI device. This means single-chip synthesizers (less oscillator and loop filter) are available now to 2 GHz. GaAs prescalers allow the classic two-chip configuration to reach speeds in excess of 5 GHz. Sophisticated digital techniques, e.g., fractional N synthesis, allow high reference frequencies for clean outputs while allowing sub-hertz frequency increments. Some of the newer transmitter/receiver techniques discussed require accurate maintenance of phase. This is one parameter that the phase-locked loop has difficulty achieving. The solution to this may well lie in the new technique of *direct digital synthesis* (DDS).

Direct Digital Synthesis

This technique generates and phase-modulates a carrier using a numerically driven or numerically controlled oscillator (NDO/NCO). The heart of a DDS is a sine/cos look-up table or waveform map containing a repertoire or menu of waveform values. (See Figure 5.13) The DDS is referenced to a clock in the same

[1] The reader is referred in particular to the *Mobile Radio Servicing Handbook* by Belcher, Fitch, Ogley, and Varrall, published by Heinneman Newes, UK.

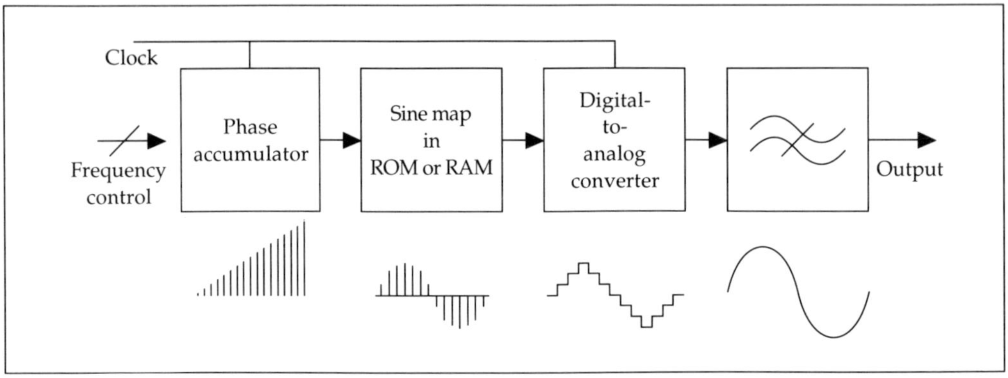

Figure 5.13

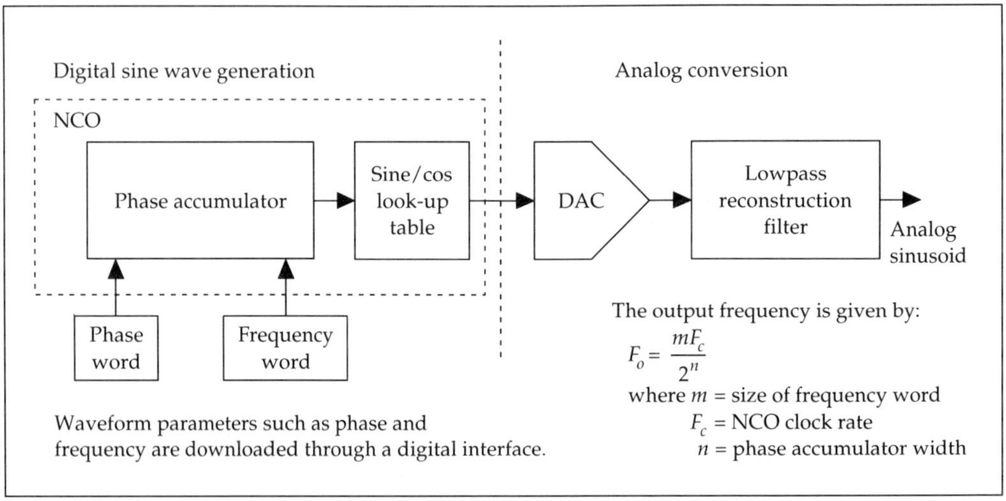

Figure 5.14

way that a phase-locked loop is referenced to a crystal. A series of addresses is delivered to the sine/cos look-up table to generate a carrier accurately phase and frequency referenced to the clock. (See Figure 5.14)

The information with respect to phase that we wish to modulate onto the carrier is generated by manipulation of the frequency word input and is a function also of the clock rate f_c. The phase of the carrier is controlled by repeatedly adding the frequency word to the contents of the accumulator. The parallel output word from the accumulator forms the waveform look-

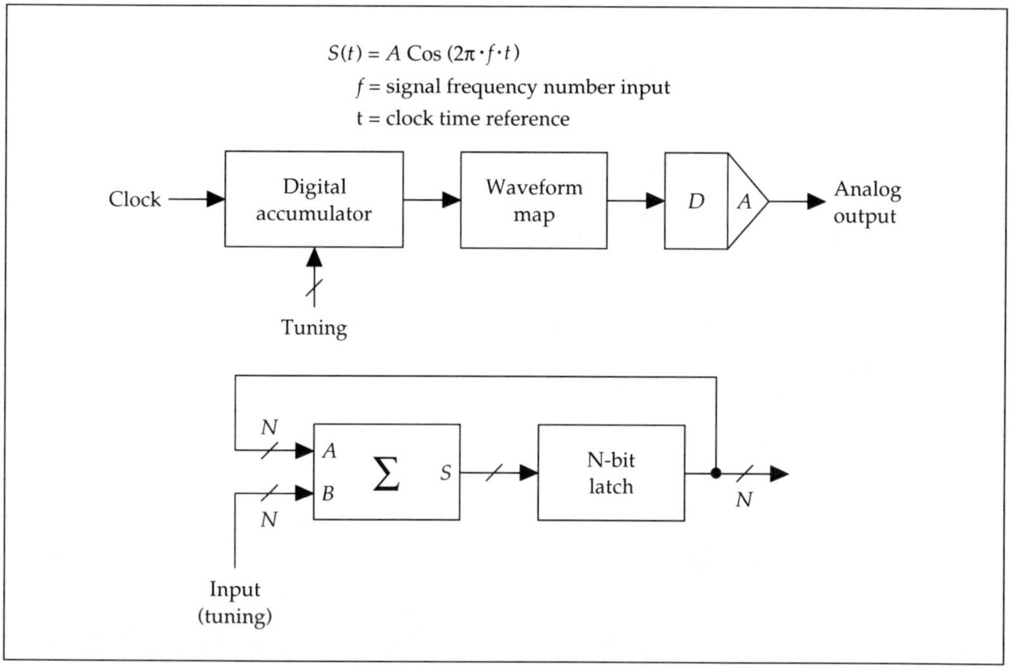

Figure 5.15

up table address. To determine the frequency word and comprehend the step size, the clock frequency and accumulator size must be known. If the accumulator is n bits long, then the maximum value that it can hold will be 2^n. In adding the frequency word to the accumulator, its overflow rate will be $f_c/2^n$. If the frequency word is equal to 1, then this value represents the lowest frequency that can be synthesized and also represents the smallest step size. (See Figure 5.15.)

Since the DDS is a traditional sampling technique, the criteria for Nyquist's sampling theorem must be obeyed. Although in theory the frequency generated can be 50% of the clock frequency, this puts practical constraints on the subsequent filtering of the reconstituted output. For this reason it is usual to work at a maximum of 40% of clock frequency.

The sine wave value of each step output by the look-up table is presented to a digital-to-analog convertor. This in turn produces an analog voltage equivalent to the digital value at its

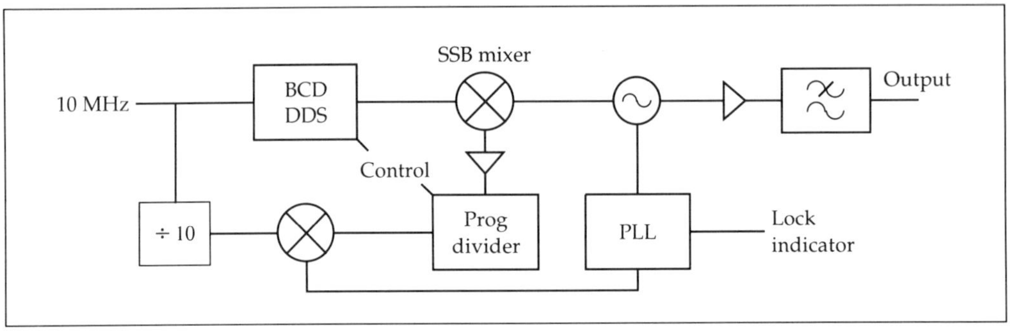

Figure 5.16

input. The quality of the D–A conversion is in practice one of the main limitations in reconstituted waveform quality.

The spectral purity of the DDS is a function of the number of bits with which the waveform is defined (similar to the 8-bit or 13-bit word length constraint in quantizing, see *Introduction*). Achieving sufficient spectral purity and working within existing digital logic and memory access speeds means that (at this writing) a DDS will not operate above 100 MHz. To operate at 900 MHz or 1.8 GHz, the DDS is nested in a conventional phase-locked loop—a complex hybrid solution but one that combines the phase accuracy of the (clock-referenced) DDS with the frequency accuracy and wide frequency range of the (crystal-controlled) phase-locked loop. The hybrid combination technique is known as direct digital interpolation (DDI). DDS and DDI techniques are being implemented currently within GSM and are the subject of pending patent applications. (See Figure 5.16.)

CHAPTER 6

CHANNEL CODING

DEFINITIONS—DETECTION, CORRECTION, PROTECTION, ERROR CONTROL

We described in our introduction the concept of source coding—the "efficient description" of our input data and the arrangement of that data into an ordered predetermined format; for example, an 8-bit word.

We now need to consider *channel coding*—the packaging, preparation, and manipulation of our source-coded input data prior to transmission.

The transmission path in mobile radio is likely to suffer from severe fading, phase cancellation (multipath effects), adjacent channel and co-channel interference, all of which will cause errors "in transit."

By adding data of a known sequence and structure to our input data we can qualify and verify the bit stream or bit block demodulated at the receiver.

If we can identify, i.e., *error detect*, a specific bit received in error in a bit block, we can *error correct*; for instance, a "0" received and identified as errored can be translated to a 1.

Errors in the mobile radio channel tend to occur in bursts. We can minimize the effect of these burst errors by interleaving—an *error protection* mechanism.

We describe the processes of error detection, error correction, and error protection as *error control*.

CODING OVERHEAD AND CODING GAIN

Because error detection and correction requires the addition of data—*coding overhead*—we must either accept a lower net data rate or an increase in bandwidth.

Error detection, error correction, and error protection (interleaving) introduce processing delay and processing complexity (i.e., cost).

We can only justify the error control "overhead" of error detection, correction, and protection if we achieve coding gain—an overall improvement in the amount of information successfully (i.e., accurately) demodulated at the receiver for a given channel condition, a specific carrier-to-interference (C/I) ratio or (a common measure) a specific bit energy-to-noise ratio (Eb/No).

We will use these benchmarks to evaluate our channel coding options and channel coding strategy for mobile radio.

ARQ AND FEC

We can define two basic methods of error control:

ARQ—Automatic Repeat Request

The determination that an error or errors have occurred somewhere in the bit block. The receiver instructs the transmitter to retransmit the whole bit block.

FEC—Forward Error Correction

The application/addition of qualifying data of a known sequence or structure to our input data.

ARQ has an advantage over FEC in that error detection requires much simpler decoding equipment and less redundancy than error correction. ARQ is adaptive in that information is only retransmitted when errors occur.

FEC is needed if a reverse channel is not available and/or if delays with ARQ are excessive. ARQ becomes very inefficient as channel bit error rate increases.

Generally it is a good principle to combine FEC and ARQ—we will look later at some specific application examples.

PARITY CODING

The simplest form of coding overhead is the *parity check* in which one extra bit is added to the data word. The receiver is instructed to expect either an odd number or even number determined by the redundant bit. For example:

Even Parity 0/1000010
Odd Parity 1/1000010

An error in the data word will change the parity allowing the receiver to detect a transmission fault and request retransmission.

The limitations are obvious. If two errors occur in the bit word, the parity check will fail to detect the errors—falsing (false validation) of the received bit word will occur.

We therefore need some mechanism to identify which specific bits in a block are received in error.

Consider for example:

	A	B	C	D
4 bits of data	1	0	0	0

The check bits are:

Rule

$E = 1 + 0 + 0 = 1$
$F = 1 + 0 + 0 = 1$
$G = 1 + 0 + 0 = 1$

The code word is therefore:

A	B	C	D	E	F	G
1	0	0	0	1	1	1

If "B" is corrupted the code word received is:

1 1 0 0 1 1 1

The check bits are used as follows:

Rule

$E(1) = A(1) + B(1) + C(0) = 2 = \text{Fail}$
$F(1) = A(1) + B(1) + D(0) = 2 = \text{Fail}$
$G(1) = A(1) + C(0) + D(0) = 1 = \text{Pass}$

Therefore

A + C + D are OK

Therefore

B is identified as error.

CODING PERFORMANCE—HAMMING DISTANCE

Because error correction is applied to blocks of data, it is often described as *block coding*. This is one form of forward error correction. We need to decide how much "extra" data or "parity bits" to add—more coding overhead delivers more error correction capability at the cost of more bandwidth.

Our measure is simply the ratio of data bits to parity bits within a bit block of defined length; e.g., a 2,1 code has a redundancy overhead of 100% and a code rate of 1/2, i.e., the bit block will require double the bandwidth of an uncoded signal to achieve a comparable net bit rate.

A 4,3 code will have a redundancy of 33% giving a code rate of 3/4 involving a bandwidth expansion of 4/3.

Now we need to evaluate the error correction capability of our coding overhead using a measurement yardstick known as *Hamming distance* (named after its author).

The Hamming distance between two code words $d(u,v)$ equals the number of components in which they differ; e.g.,

$$u = 1\ 0\ 0\ 1\ 0\ 1\ 1\ 0\ 0\ 0\ 1$$
$$v = 1\ 1\ 0\ 0\ 1\ 0\ 1\ 0\ 1\ 0\ 1$$
$$d(u,v) = 5$$

The minimum distance of a code equals the minimum Hamming distance between code words in the code.

A code with minimum distance D_{min} has error-correction-detection capability $(D_{min} - 1)/2$ or error-detection capability $(D_{min} - 1)$.

Hamming's work in the early 1950s on coding theory (and Shannon's work in the same period) has become the basis for a whole generation of codes of varying length and structure developed to meet specific error correction requirements. Most codes are named after their authors. For example:

Golay Code and Extended Golay

Golay is a 23,12 code. Extended Golay is a 24,12 code constructed by adding an overall parity bit. Extended Golay is more powerful than a standard Hamming code but requires a more complex decoder.

Cyclic Codes

Many codes have a cyclic structure. Additional parity bits are added to a bit block to provide a "first line of defense" validation check. If the bit block fails the parity check, the coding system can either request retransmission of the block or ignore the information in the block altogether, using the information from the prior bit block to fill the information gap. This works particularly well when coding the output from speech codecs with a high degree of sample-to-sample correlation. Bit blocks that are validated are passed on to a second stage of coding (for example, convolutional coding), then interleaved, then modulated onto a carrier for transmission.

BCH Codes

A group of block codes authored in 1960 by Bose, Chaudhuri, Hocquenghem with a repertoire of block lengths, code rates, and error correction capability.

At a block length of a few hundred bits, BCH codes are considered to outperform all other codes with the same block length and code rate.

Reed Solomon Codes

A special subclass of BCH codes often used for burst error correction.

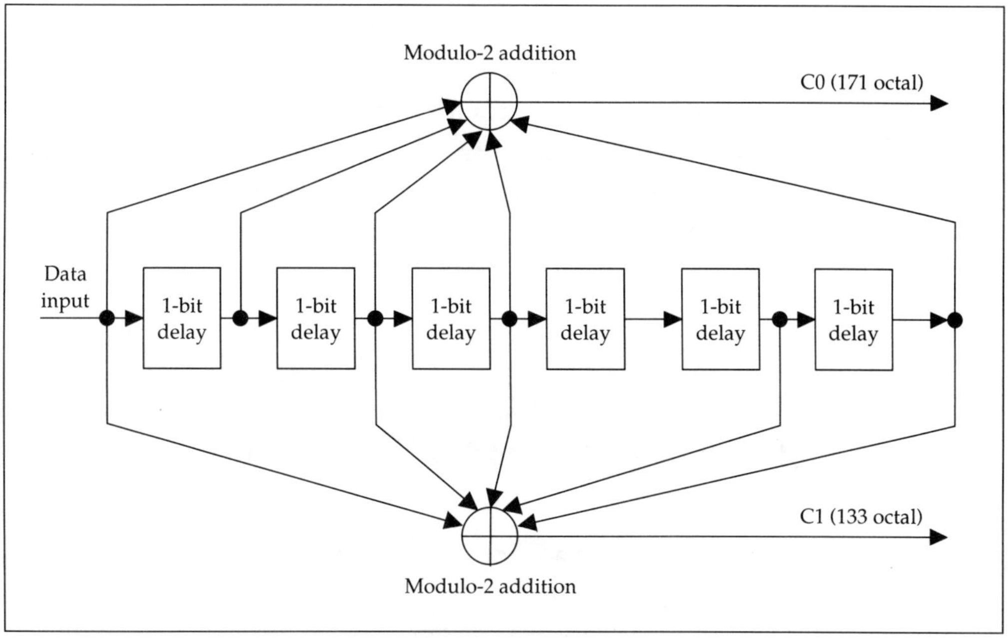

Figure 6.1

Convolutional Codes

In convolutional coding a single input bit value is combined with—for example—six previous input bits to produce two output values.

Two bits are output for every 1-bit input, therefore this is a 1/2 code, but note some of the validation bits are also information bits used as check bits because they have already been verified. (See Figure 6.1.)

CONSTRAINT LENGTH

The error correction capability and complexity trade off is expressed by the term *constraint length*. Constraint length is determined by the number of input bits maintained in the memory of the encoder; e.g., for the convolutional code above, constraint length (K) is 6.

Error correction capability increases with increasing constraint length. Increasing code rates result in reduced bandwidth efficiency. Increasing constraint lengths require exponentially increasing computation complexity in the decoder.

A code of length 6 or 7 is a good compromise.

COMPLEXITY TRADE-OFFS—THE VITERBI EXAMPLE

We can say as a general rule that decoders are more complicated than encoders.

The most popular decoding technique uses a *Viterbi algorithm*—conceptualized by Viterbi in the 1960s, functionally realized first in the early 1970s.

The Viterbi algorithm is a maximum likelihood decoding technique, assessing the likelihood of each of the possible states of the encoder for each encoder output group, making decisions as to the probable encoder input; e.g., for our convolutionally coded example:

> *Our six delay elements imply a total of 64 binary path decisions that need to be computed and evaluated for each data bit.*

A Viterbi decoder therefore consists of two sections: decision computation elements—(performing addition, comparison, selection) and a decision memory structure.

CONCATENATED CODES

Our last group of codes or combinations of codes to consider are *concatenated codes.*

A concatenated code uses two levels of coding; an inner code and an outer code. The inner code is usually configured to correct most of the channel errors. The outer code, usually a higher rate (lower redundancy) code, then reduces the probability of error to the specific level.

Concatenated code achieves a low error rate with an overall implementation complexity less than that which would be required by a single coding operation.

CODING GAIN COMPARISONS

We can now evaluate some coding options in terms of their coding gain. Coding gain is defined as the reduction in the required

$$\frac{E_b}{N_o}$$

to achieve a specific error probability of the coded system over an uncoded system with the same modulation and channel characteristics. For an example, see Figure 6.2.

We can take three practical examples of channel coding.

GSM as an Example

The speech codec in GSM outputs information (the description of the analog speech waveform) in 260 bit blocks.

The 260 bits are delivered to the channel coder in decreasing order of importance, the 182 class 1 bits first.

The first 50 class 1 bits have three parity bits added to them using a cyclic redundancy check code. This is a classic application example of cyclic coding (see our earlier notes in this section). If the 50 class 1 bits fail the parity check, they are ignored and replaced by the 50 bits from the prior bit block.

The total bits (now 185 with 4 tail bits, all equal to zero), are added to give a *total 189 bits*. These 189 bits are then encoded using 1/2 convolutional code to give 378 bits.

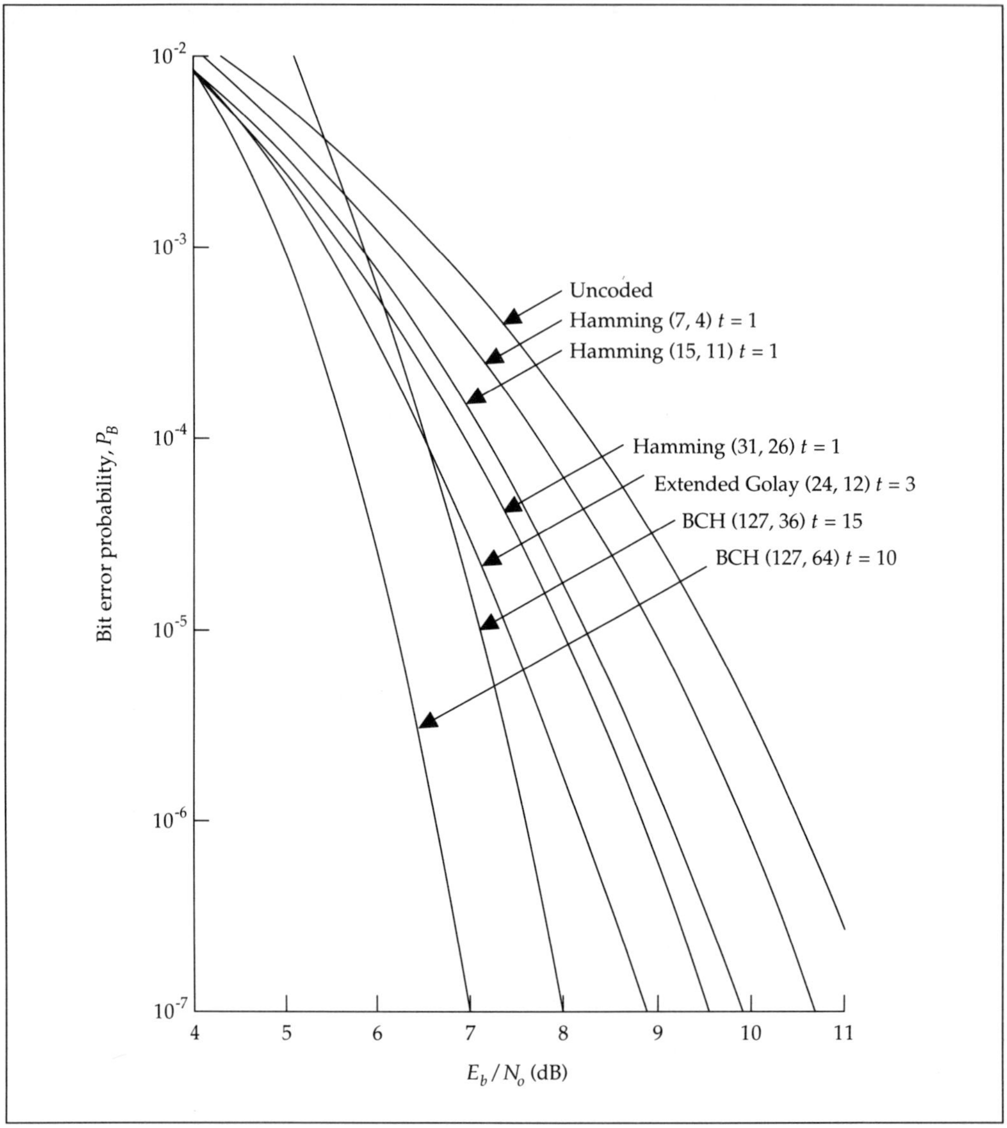

Figure 6.2 *P_B versus E_b/N_o for coherently demodulated BPSK over a Gaussian channel for several block codes and cyclic codes*

Seventy-eight class 2 bits, which do not require error protection, are added to 386 bits to give an output of 465 bits in a 20 ms duration, making up a gross coded bit rate of 22.8 kb/s. (See Figure 6.3.)

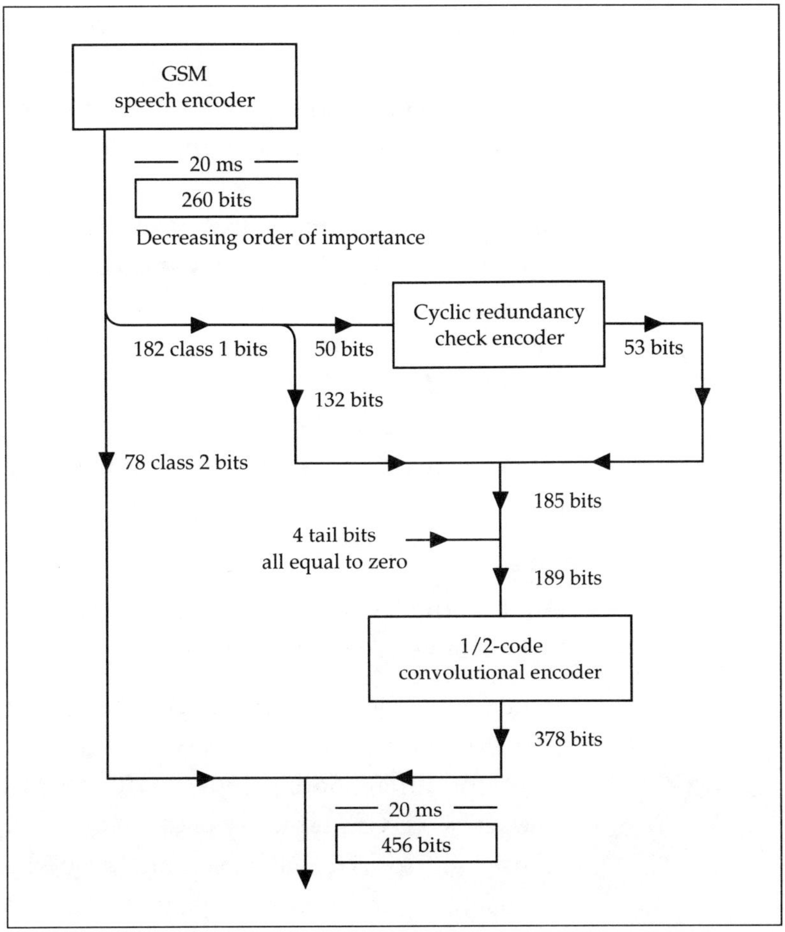

Figure 6.3

Coding performance can be improved by implementing soft decision decoding using a combination of hard decision digits and confidence digits. A proposal for GSM is to link a confidence threshold decoding mechanism with the output from the channel equalizer (measuring in real time the transmission channel quality characteristic).

CT2 as an Example

CT2 operates in a less onerous propagation environment, compared to GSM—a short path and a user (normally pedestrian), who will be more or less stationary; i.e., less dynamic multipath signal/phase cancellation.

The objective in any network—GSM or CT2—is to provide plenty of error protection for system signaling; undetected errors within the signaling bit stream will seriously disrupt the call set up and call maintenance process.

In CT2, a 48, 64 BCH code is used to protect signaling messages made up of 48 data bits, 15 cyclic redundancy check bits, and 1 extra bit to provide an even bit parity check on the whole 64 bit word.

Up to five bit errors can be detected—a good compromise between implementation complexity and error detection performance.

Digital Squelch as an Example (Golay Code) and the Need to Minimize Frequency Spread

Digital controlled squelch (DCS) is being introduced to replace CTCSS (continuously controlled sub-audio squelch) as a mechanism for differentiating between different user groups sharing a network. CTCSS typically uses 16 tones, which identify up to 16 user groups. DCS achieves the same identification process but using a continuously repeated 23-bit word, with a 23/11 Golay code—23 transmitted bits, 12 data bits, with 3 bits used to initiate the Golay sequence, leaving 9 information bits. A 9-bit word gives $2^9 = 512$ possible codes. For reasons which will shortly become clear, all codes with seven or more 1s or 0s in sequence are discarded, resulting in 104 "clean" codes. DCS has to coexist with the audio passband—it is present continuously while speech is present, preventing other users from accessing/listening to the channel unless they are using an identical code sequence (i.e., as with CTCSS, only valid mobiles will have their "squelch" lifted to allow conversation to take place). CTCSS tone sequences range from 67 to 250 Hz—comfortably below the audio passband. With DCS, the bit rate is 134 bits per second. The highest fundamental frequency is half the bit rate, i.e., 67 Hz. The lowest fundamental frequency is produced when six 0s follow six 1s and is equivalent to one-sixth of the maximum fundamental frequency, i.e., 11.17 Hz. The third harmonic of 67 Hz falls at 201 Hz, the fifth harmonic at 335 Hz, the seventh harmonic at 469 Hz. The seventh harmonic is in the audio passband but is three octaves from the fundamental, therefore sufficiently suppressed.

Squelch tail elimination, achieved in CTCSS by a phase reversal of the audio tone, is accomplished in DCS by sending bit reversals at twice the normal bit rate, i.e., 134 Hz.

Our final topic—interleaving (error protection)—we will address in the next chapter.

NOTE Readers wishing to study error control mechanisms in more detail are referred to the excellent book *Digital Communications, Fundamentals and Applications* by Bernard Sklar, published by Prentice Hall.

NOTE The concept of using a sub-audio low bit rate digital bit stream as an "identifier" is also used in N-TACS and N-AMPS. See section on digitally enhanced analog cellular systems in Chapter 9.

CHAPTER

7

DATA OVER RADIO APPLICATIONS

DEFINITIONS AND DELIVERY OPTIONS

We have covered the basic techniques used in mobile radio data applications. Now we need to consider some practical applications.

It is generally considered that non-speech traffic (e.g., fax machine to fax machine, computer to computer, or numeric/text messaging) will increase as a percentage of overall telecommunications traffic. In the PSTN, non-speech traffic now constitutes over 10% of total revenue.

Within mobile radio, there are other reasons for using data over radio. Text messaging increases channel utilization. Typically the "information" contained in a 30-second voice conversation can be despatched as an ASCII-based text message using 1200 baud rate FFSK and requiring less than three seconds of air time.

For speech bearers in mobile radio, speech digitization is used to reduce bandwidth, improve performance in poor carrier-to-interference conditions and correct for multipath effects.

Our delivery options are either fixed-point-to-fixed-point—for example (outdoor) microwave links, telemetry or telecommand, or (indoor) electronic point of sale (EPOS), or fixed-point-to-mobile or mobile-to-mobile which we define as:

Private mobile radio—closed user groups
Trunking
Cellular
Paging
Cordless outdoor
Cordless indoor—cordless local area networks.

EXISTING BANDWIDTH CONSTRAINTS

Our operational constraints include the radio environment. For example, constraints include established channel spacing, 30 kHz, 25 kHz, 12.5 kHz optimized for analog speech; or existing system limitations, for example, delivering FFSK down an rf processing chain with limiting and filtering circuits designed for speech. Mismatch in a filter chain, causing differential delay, could be virtually unnoticeable when processing analog speech but significantly disruptive to data.

MOBILE PROPAGATION CONSTRAINTS

It is worth reminding ourselves of the different nature of the "product" we are attempting to deliver. Analog is a smooth, continuously varying waveform, which can be sympathetically AM- or FM-modulated onto a sinusoidal rf carrier. Digital consists of discrete symbols, which (typically in mobile radio) will be frequency- or phase-modulated onto the carrier.

Our carrier and associated modulation (in whatever format) will be influenced (generally degraded) by our propagation environment.

We can define our possible propagation paths as:

Direct path—clear from terrain contour

Line of sight—clear from buildings

Out of sight—ground reflected and building reflected (typical of the mobile urban environment)

Mobile propagation constraints include:

Long-term fading—determined by base-to-mobile terrain contour.

Short-term/multipath/Rayleigh fading—determined by reflections from buildings and structures.

Doppler effects—determined by the movement of the user by car, train, or on foot, through the multipath environment.

Rayleigh fading is exhibited on our radio signal as flutter. At 900 MHz, flutter patterns have a shorter spatial wavelength and are affected by "street furniture"—900 MHz wavelength is 330 mm.

The effect, passing through an urban environment at, for example, 20 miles per hour, is a rapid nulling of the signal level—up to 20 nulls per second with a 40 dB fluctuation from 10 dB above to 30 dB below the mean signal level. (See Figure 7.1.)

Rayleigh fading can cause unwanted phase and amplitude modulation and can cause error rates as high as one bit in 50, which is 2000 times worse than a "poor" telephone channel.

As frequency increases, attenuation increases and electrical noise decreases. Noise is more important than attentuation at VHF (70/175 MHz) and UHF (425/465 MHz).

Cellular at 900 MHz has the advantage of a relatively low noise floor (though typically high levels of adjacent and co-

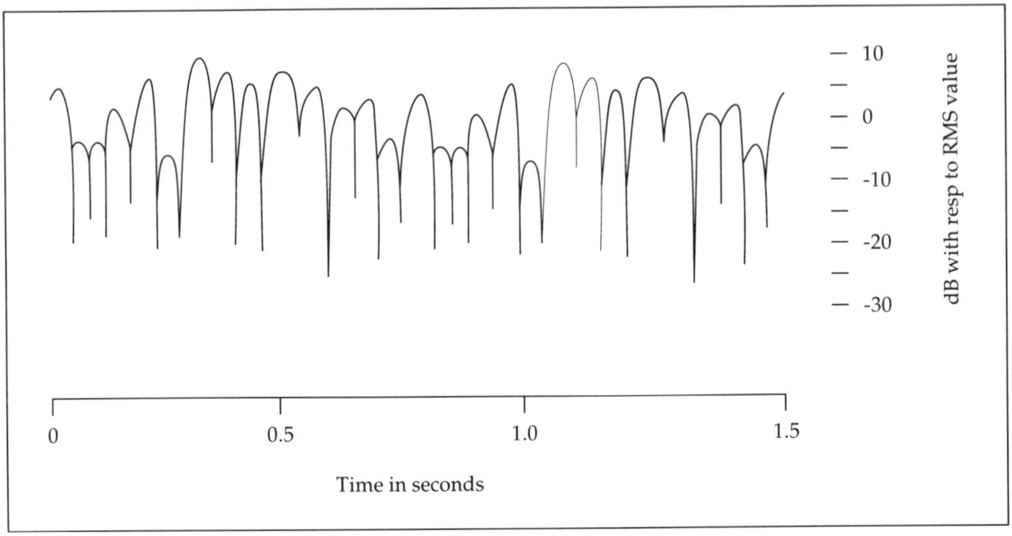

Figure 7.1

channel interference), and the disadvantage of severe multipath signal phase cancellation.

As we move higher in frequency, to microwave and millimetric frequencies, the bandwidth available increases (good for high data rate transfer applications), and attenuation increases (free space, weather, and resonant effects). We examine millimetric broadband data bearer applications in Chapter 14.

SYSTEM ARCHITECTURE LIMITATIONS

Existing radio system architectures, designed for speech are, not surprisingly, not always optimal when used as data bearers.

Trunking and cellular systems use signaling channels to set up traffic channel pairs (simplex or semi-duplex press-to-talk for trunking, full duplex for cellular). The call set-up process can take 15 to 20 seconds—relatively wasteful when the end result is to send a three-second burst of data!

In-band signaling, for example in AMPS or TACS cellular with a mid-sentence 300 m/s burst of 10-kb or 8-kb signaling information can be seriously disruptive to data.

Quasi-synchronous systems, particularly AM, designed for analog speech may have frequency and phase translations occurring between sites—tolerable for analog speech but seriously disruptive to data.

Simpler network architectures—for example private mobile radio networks with 25 kHz or 12.5 kHz spacing—are therefore quite well suited to become shared or dedicated data bearers using, for example, 1200 baud rate FFSK.

CODING FOR TWO-WAY RADIO

Some error coding specifically designed for the mobile radio propagation environment is needed. An example of one proprietary United Kingdom protocol is to use a synchronous shortened BCH 14,10 code (14 data bits, 10 check bits, resulting in 2-bit error correction), with a 192-bit block (8×24), then interleaved over eight subsequent blocks, delivering typically a net data rate of 100 characters per second.

DATA OVER TRUNKED SYSTEMS

Data over trunked radio presents us with three options. The 1200 baud FFSK bit stream used in the signaling channels is generated by an internal modem. This modem can be used to send short data messages over the signaling (control) channel—a short data message of 184 bits of free format data, with an address codeword (head) and four data codewords. The network must maintain some discipline over how much use is made of the signaling channels. If the signaling channels are fully occupied with status messaging, there will be no mecha-

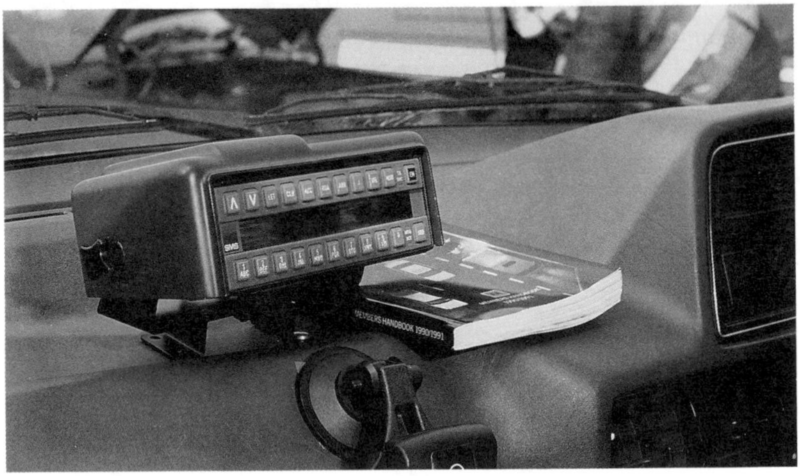

Figure 7.2 *Dash-mounted data input terminal. Note specially designed cowling to minimize heat gain through vehicle windscreen.*

Figure 7.3 *Dash-mounted mobile data terminal (MDT) with full Qwerty keyboard and windscreen-mounted liquid crystal display with graphics capability.*

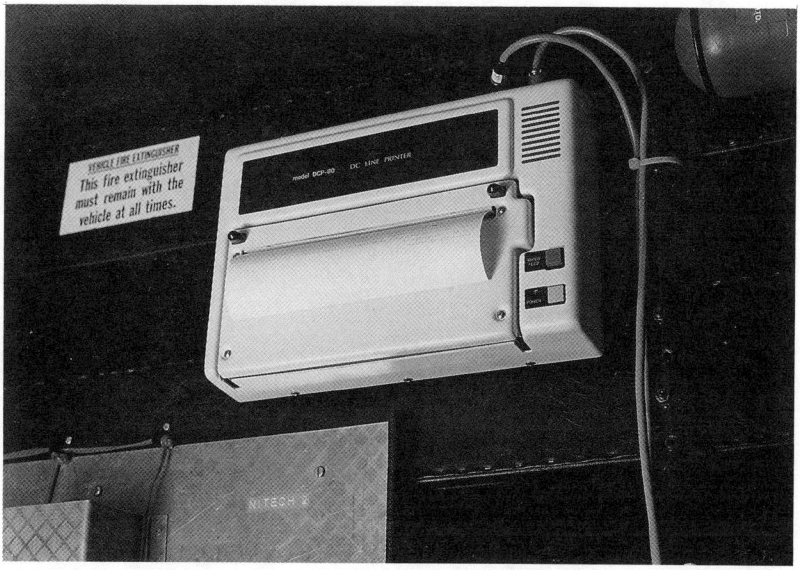

Figure 7.4 *Bulkhead-mounted printer*

nism for controlling the network! The modem uses a 63/48 cyclic code.

An alternative is to use the modem to send data over a traffic channel. The third option is to ignore the internal modem and address the traffic channel with a separate modem (preferably with mobile radio specific error control). The called party will need a modem with the same error control protocol.

A typical and appropriate application for data over closed user group networks, or trunking networks individually serving closed user groups, is digitally assisted dispatch. Two examples in the United Kingdom include the Royal Automobile Club and British Gas (North Thames). The RAC CARS (Computed Aided Recovery Service) project uses 120 radio base stations nationwide to deliver data messages to 1500 patrol cars supported on five VHF low band 12.5 kHz channels, with a 1200 baud rate address/status messaging service, and a 2400 baud rate service for text on the traffic channel.

British Gas North Thames will be serving 11 district offices, 550 vehicles with X25 network capability.

NOTE Considerable effort has gone into the interface between relatively low bit rate mobile radio networks and X25. X25 was designed to optimize 64 kb/s transmission over poor-quality telephone land lines and has, for that reason, an error correction overhead that is far too high for today's "clear" 2-megabit links. There will always be a substantial rate adaption process between mobile radio networks and the land line network.

NOTE United Kingdom and European data networks are dwarfed by the SMR (Specialized Mobile Radio) networks in the United States and by dedicated networks such as Ardis, the network set up by Motorola and IBM in 1985 across the United States to provide communication to IBM's 20,000 service engineers.

DATA OVER ANALOG CELLULAR

Analog cellular has a wider bandwidth than most PMR systems (25 kHz versus 12.5 kHz in the United Kingdom), a low noise but high interference propagation environment and multipath effects—good news and bad news for data transfer.

Vodafone/Vodata have implemented a nonproprietary[1] error control protocol—CDLC, cellular data link control, to take account of the mobile radio channel characteristics and the dynamically changing network of the channel. Figure 7.5 shows the Vodata CDLC modem.

CDLC provides a progressive "line of defense" to protect the data as channel quality deteriorates (i.e., as bit error rate increases).

CDLC first defines a frame structure (See Table 7.1). A 48-bit leading synchronization sequence defining the frame length, an address field managing the routing of the frame, a

[1] (Nonproprietary means that Vodafone/Vodata encourage other networks to adopt the protocol to develop a common standard for error control across a number of systems).

Figure 7.5 *Vodata CDLC modem*

Table 7.1 *The CDLC frame structure before FEC and bit interleaving*

48-bit leading synchronization	8-bit address field	8-bit control field	Information field 8N bits up to 504 bits)	16-bit fraime check sequence

control field sequencing the forward and reverse flow of information for triggering ARQ, the actual information (up to 504 bits) and a 16-bit frame check sequence.

The protocol uses two levels of coding; BCH coding with 100% coding overhead for high bit error rate channel conditions, and a lighter weight RS code for (relatively) low bit error rate channel conditions.

First line of defense is therefore the RS code—four additional parity bytes (symbols) added to a frame will correct any two 8-bit symbols in error—an adequate error control mechanism for channel error rates up to one in 200.

BCH coding (the second line of defense) takes each multiple of eight bits within the CDLC frame and codes the eight bits into a 16-bit codeword. BCH coding allows up to two bits to be corrected per codeword, the identification but not correction of three to five bit errors. Six bits in error may not be detected by BCH but the errors will show up as an invalid frame check sequence, which will trigger retransmission (ARQ—the third line of defense).

The fourth line of defense is interleaving. In multipath fading, bit errors will tend to occur consecutively causing our BCH-coded FEC (forward error correction) to fail and trigger ARQ—substantially slowing data transfer. We want to minimize the number of times that ARQ is triggered.

The aim of interleaving is to split up errors so only one or two errors occur in any one codeword. For example: Our bit stream contains 16 codewords, each codeword contains 16 bits. Bit interleaving creates 16 new codewords. The first codeword contains Bit 1 of each of the original codewords. The second codeword contains Bit 2 of each of the original codewords, etc. (See Figure 7.6.)

Finally, if all else fails, the fifth line of defense is to buffer the data until transmission conditions improve.

Combining BCH-coded FEC and interleaving in channel conditions with bit error rates at 1 in 50, a frame length of 1000 bits will have all errors removed with only 10% of frames requiring retransmission.

Vodafone offers a "managed access conversion service" (V-MACS), which means that data from a CDLC modem delivered to the network will be delivered through the network. Prior to delivery into the PSTN, the data will have the CDLC error control removed. The data can then be delivered through the PSTN to any PSTN-compatible modem; i.e., the called party does not have to own a CDLC modem. Figure 7.7 demonstrates access to the Vodafone cellular network using the CDLC modem and cellphone.

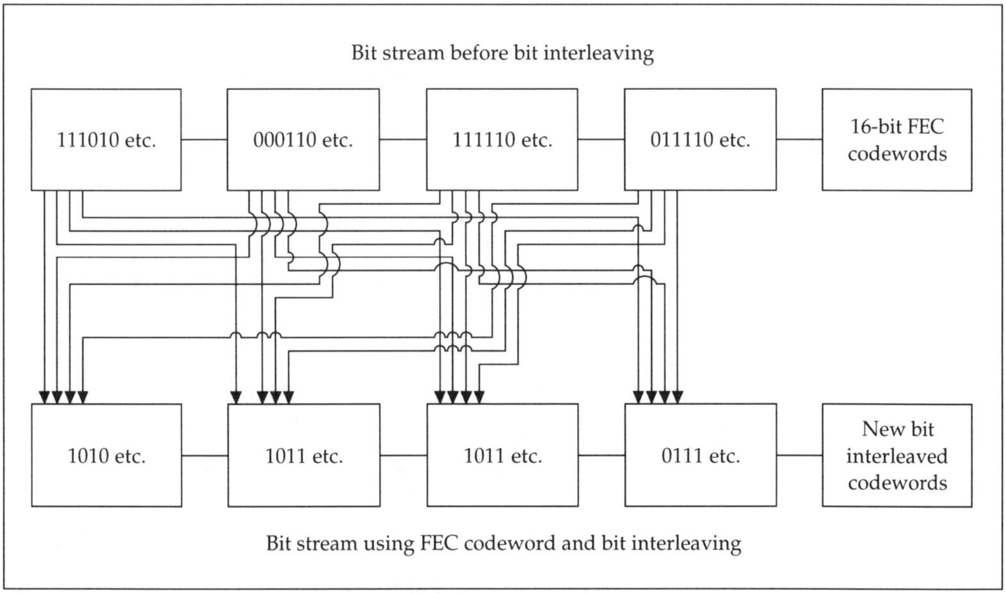

Figure 7.6

Figure 7.7 *Access to Vodafone cellular network using CDLC modem and Cell-phone.*

The same service can be provided to large corporate users with direct access to the Vodafone network (i.e., the data never enters the PSTN but is delivered directly through the corporate PBX to standard office modems).

The user has the choice of a two-wire circuit with a slow reverse channel (150 bps) for ARQ control. This will deliver up to 2200 bps or a four-wire cellular fax, which can deliver up to 9600 bps with a claimed realistic net data transfer rate of 2400 bps at 70 mph.

Users of Vodata in the United Kingdom include the National Rivers Authority, British Rail (a telemetry application), and the Metropolitan Wheel Clamping Operation.

Cellnet opts to make its network "transparent" to data users; i.e., the user takes the responsibility of deciding on a modem, preferably using CDLC or similar mobile radio specific protocol, delivers the data to Cellnet which delivers on to the called party who must have a terminal with the same protocol.

In the United States, trial project demonstrations have shown the practicality of implementing a packet data system—3 kHz data channels interleaving into the guard bands either side of the existing AMPS speech channels, delivering a net data rate of 2400 bps. The concept is to insert narrow bandwidth data traffic channels in the "lightly modulated" spaces between voice traffic channels; e.g., those channels in the radio plan that would fail to meet adjacent channel interference crtiteria using voice can be used to deliver data traffic provided bandwidth is constrained to meet system C/I (carrier-to-interference) criteria. See Figure 7.8.

DEDICATED PUBLIC DATA NETWORKS

Which brings us to the transatlantic phenomena of dedicated public data networks. In the United Kingdom there are four licensees tasked to provide mobile data services nationwide;

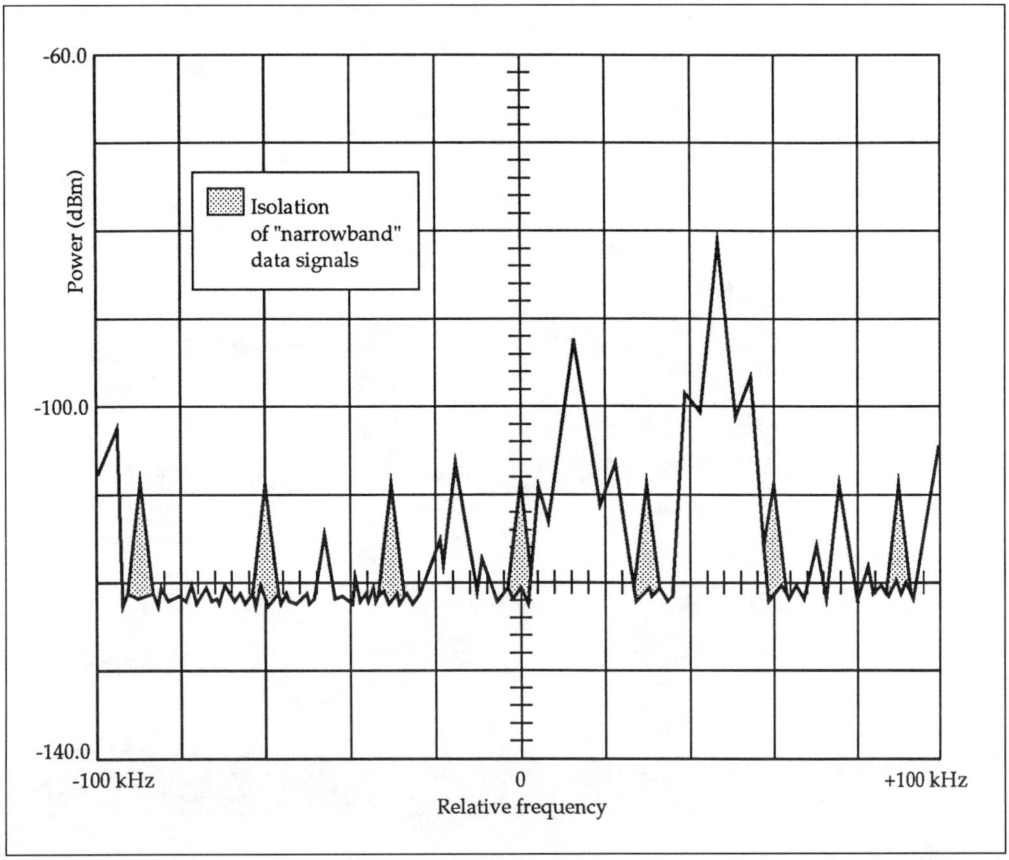

Figure 7.8

Hutchison (using MDI/Motorola technology), RAM (using Mobitex/Ericsson technology), Cognito (using proprietary technology), and Motorola (using USA/IBM Ardis network technology). Paknet (a Racal/Chubb/Mercury joint venture) is licensed to offer a fixed radio-based service targeted at EPOS (electronic point of sale) applications offering wireless validation of credit cards in (Paknet claims) three to five seconds—twice as fast as a dedicated fixed link, four times as fast as a dial-up link (see Figure 7.9). The Paknet network is an "underlay" under the Vodafone network, sharing base station sites but operating at 160 MHz (transmit 159.7 to 159.9 MHz, receiver 164.2 to 164.4 MHz), with 14 duplex forward and reverse 12.5-

Figure 7.9 *Electronic point of sale application using access to Paknet Network for fast off-air validation of credit card/charge card transactions via Paknet "Radio Pad."*

kHz channels and two guard channels (see Figure 7.10). Data rate is 4800 bps, using packets of 128 bytes interleaved onto a 27 ms time slot.

NOTE The Paknet network and operational protocol will support mobile applications but this would require a variation to their fixed operating license, which is under discussion.

Three of the mobile data networks (Hutchison, RAM, Motorola) are at UHF; Cognito is in the lower sub-band of Band III—189 MHz—(Band III is the frequency sector vacated by the United Kingdom's discontinued, 405-line black-and-white television service). See Figure 7.11

Figure 7.10 *Paknet base station. A telemetry outstation, remote site. Note optional integral battery.*

Cognito supports a 6.1 kb/s data rate on 12.5-kHz channels. The base stations (full duplex) manage communications between the network and all the mobiles (two frequency simplex) within their area with a contention protocol to manage the single up-link frequency in each cell.

When a mobile registers, its location is registered through the base station onto a directory server. A broadcast channel

Figure 7.11 *Cognito "Messager" hand-held portable terminal*

from the base to all mobiles downloads information on new frequencies in neighboring cells and an absolute time reference.

The network uses time division multiple access with 12 slot frames; each slot has 768 bits with a slot length of 0.125 seconds (at 6.144 kb/s). Mobile transmission slots are separated by guard periods of 36 bits. (See Figure 7.12.)

Network architecture is optimized to reduce the power consumption of the terminals. A mobile terminal is alerted to a

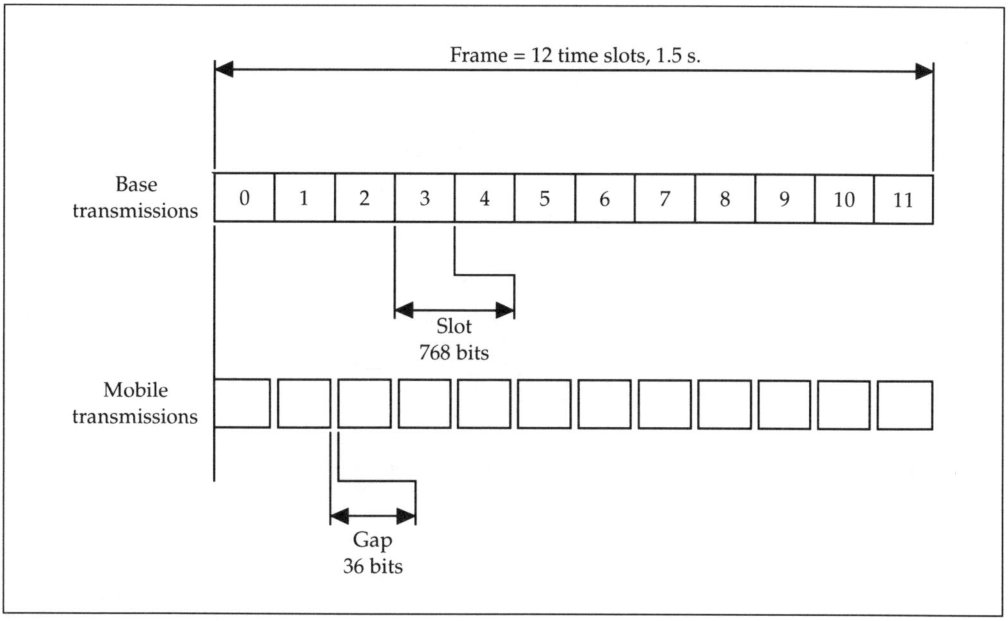

Figure 7.12

packet due for delivery by a down set-up message transmitted in a specific time slot. The mobile only powers up to monitor a predefined slot.

An optional "very low consumption" mode involves the network informing the mobile of the current length of a multiframe and in which frame of the multiframe the mobile should listen to receive a down-link message—trading a delay in message receipt for low standby power consumption.

The RAM network uses Ericsson technology and the Mobitex standard—now administered by an independent Mobitex Users Association.

The "RAM Plan" is to provide the United Kingdom with nationwide coverage from 200 base stations. Early pilot schemes included Heathrow, Gatwick, Crystal Palace, and Bolton. A subscriber's host computer accesses a local switch node, the local switch node supports up to eight base stations plus direct lines to 20 or more subscribers. The network has a mailbox storage facility (store and forward) and normally delivers

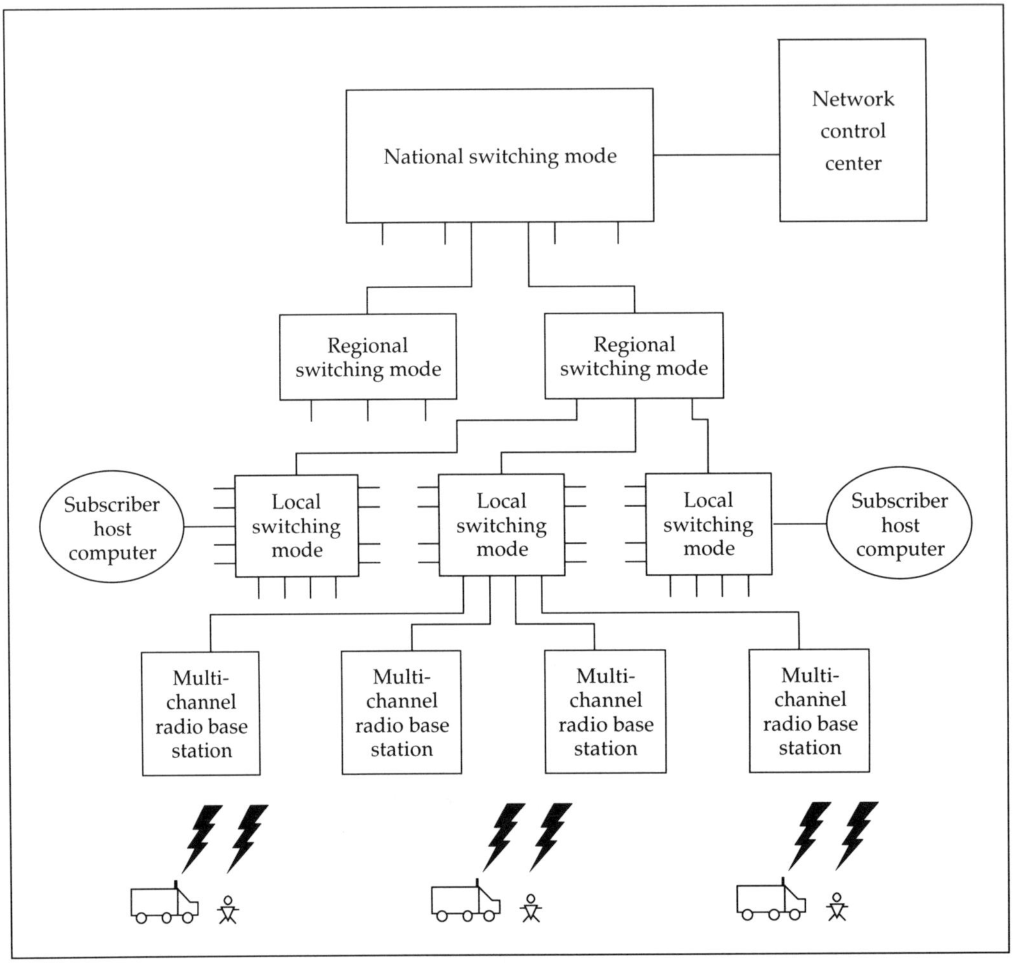

Figure 7.13

data packets of between 18 and 512 characters, or multiple packets. (See Figure 7.13.)

Mobitex-based networks, dating back to the early 1980s now have 15,000 to 20,000 terminals in use in Scandinavia (Sweden, Norway, Finland), and (now) Canada, the United States and the United Kingdom.

Initial application experience involved some EMC (electromagnetic compatibility) problems with the equipment (unwanted crosstalk between radio, modem, and printer) and some difficulty with automotive integration—unreliability caused by the high temperature and humidity gradients typically encountered in a car, especially in Scandinavia (–30° C to +40° C)!

The Mobitex protocol is claimed to be able to support up to 1000 users per channel, optimized to the "most common message length" (40 to 50 characters). The protocol is based on individually addressed numbered packets. A lost packet can be retransmitted and reassembled sequentially at the receiver.

All mobiles listen to the base station but only respond to traffic with a recognizable address—similar to mobiles in a two-way speech mobile radio network only responding (lifting squelch) to a specific CTCSS (sub-audio) tone or two- or five-tone in-band tone sequence.

The Mobitex mobiles therefore cannot hear one another. Transmissions can collide so the base station has to manage access to the system.

The networks worldwide have matured from 1200 baud rate systems, to 4800 to 8000 baud rate at 80, 160, 400 and 900 MHz.

Mobile terminals are available from a number of manufacturers including Mobira, Mitsubishi, Ericsson, and Philips.

Dedicated Mobile Networks Compared

Cognito	6.144 kb/s 3-level FSK 32 channel pairs 5-watt mobile power
Mobitex	8.000 kb/s 3-level FSK 20 channel pairs 15–20 watts
Motorola	4-level FSK 16 channel pairs 4 and 20 watts

Pan-European Standards

The ETSI RES 6 Committee is working toward a common Pan-European standard for dedicated mobile radio networks with the target of supporting all X25 network interfaces and delivering 128-byte packets in 100 ms.

There is a strong lobby for a Pan-European frequency allocation—a 4-MHz segment of spectrum by 1995, 10 MHz of spectrum by the year 2000.

CHAPTER

8

PAN-EUROPEAN DIGITAL CELLULAR—GSM

POLITICAL RATIONALE

GSM stands for Global System Mobile and the specification of GSM has been the responsibility of the Groupe Speciale Mobile Committee. GSM is a product of the specification industry—one of the fastest growth industries in the world today. For GSM, the technology is expected to deliver a political benefit—the unification of Europe through a common mobile communications network.

TECHNICAL AND COMMERCIAL RATIONALE

Fortunately there are other more prosaic motivations driving the GSM implementation program. The motivation of the manufacturer and network operator has been and must be to improve margins and fiscal return on investment by reducing costs—the per-subscriber cost of providing network infrastructure; i.e., the radio engineering, network engineering, and software engineering costs, and the costs of manufacturing mobile phones. If the total network cost-per-subscriber can be reduced, the cost of providing air time is reduced. The network operator buys competitive advantage. Lower costs can either be passed on to the user or retained as added margin.

Some operational benefit to the user can also be translated into a market advantage (same pricing for better service) or increased margin (higher terminal costs and/or higher tariffed air time).

The use of sub-micron technology and high-level circuit integration combined with a time-division multiple-access structure, in which the mobile is only transmitting one time slot every 8 slot frame, promises a lower power budget and a better talk-time-to-weight-to-size ratio than analog phones offer.

GSM technology is designed to reduce the effects of adjacent channel and co-channel interference, improve power control and hand-off, and reduce the effect of multipath signal phase cancellation. This should improve the up-link and down-link "quality" of service to and from the mobile—improved access (less blocking) and fewer mid-conversation dropped calls.

GSM also provides extra ISDN-type facilities such as calling line identification, a (relative to TACS) high bit rate data bearer, Pan-European roaming, and Smart Card enabling and billing.

The technology and fiscal mix only makes sense if large numbers of users can be encouraged to use the network. A GSM base station will be more expensive than its analog counterpart. It is only less expensive per subscriber if it can be loaded close to its full capacity, with a capacity per cell site two to three times greater than an analog cell. rf engineering in GSM will still represent over 60% of the cost of the network.

HISTORY

Digital cellular options began to be discussed seriously in the late 1970s; i.e., before the analog networks were introduced. Early contenders for a "European" system included a Philips proposal, MATS-D, an asymmetric system using TDMA only on the down-link (base to mobile), a broadband proposal (CD900), and various intermediate ideas.

The GSM committee was established in 1982 to preside over a politically and technically acceptable compromise and, 10 years later, it is arriving.

The consequence of a consensus approach to specifying the network is a wide applications requirement. The Nordic delegation to the committee insisted that the network should support macrocells with a 35-km radius to service Scandinavia's "wide open spaces." The British insisted the network should support microcells with a radius of 200 metres. The Germans insisted the network should support "in-car communication" at up to 250 km/h, the French insisted the network should support communication to and from their TGV (*grand vitesse*) trains. The British would have preferred the network to be optimized for their (virtually stationary) pedestrian user community using low-power, hand-held microportables.

CONSEQUENCE OF WIDE APPLICATION REQUIREMENT

The wider the application requirement, the wider the performance requirement. Large cells mean long path-link delays. Mobiles must be instructed to transmit in advance of the base station's reference signal—a timing advance of up to 63 bit periods, a long synchronization sequence, and frequency correction.

A mix of large cells and small cells requires a wide dynamic range of power outputs; 2.5-watt to 300-watt base stations with eight (3 dB) attenuation levels, class 1 mobiles from one to 20 watts, class 5 mobiles from 0.8 watts down to 0.2 watts.

High-speed operation implies Doppler effects and a rapidly changing multipath environment. A mobile can traverse a whole wavelength within a time slot. This places severe demands on the channel equalizer—computationally expensive, power hungry and slow (5 ms or so). Computation complexity involves cost and processing delay. In GSM, the codec, the channel coder, the multiple-access structure (interleaving), the radio path, and (in the receiver) the channel equalizer, all contribute to the "delay budget." It has been one of the design challenges of GSM to keep this delay below a point (about 70 ms) at which it becomes noticeable or irritating to the user.

NOTE Geostationary satellite links have delays of up to 260 ms but, as any user of a satellite link will testify, this is severely disruptive to "conversation comfort."

It was also decided to retain the 45-MHz duplex spacing between transmit and receive channels implemented in analog TACS, with the anticipated need to introduce GSM as a parallel system. Duplex spacing means a duplex filter—one of the harder parts of an rf front end to get right. Duplex filters today have been miniaturized but are still power hungry (up to 3 dB loss) and retained in the radio only as a necessary evil.

GSM SPECIFICATION

It would be impractical to reproduce 15,000 pages of GSM specification verbatim, so here are some edited highlights. (If you *do* want to read the whole specification, it can be purchased from the European Telecommunications Standards Institute (ETSI) in Nice, France.

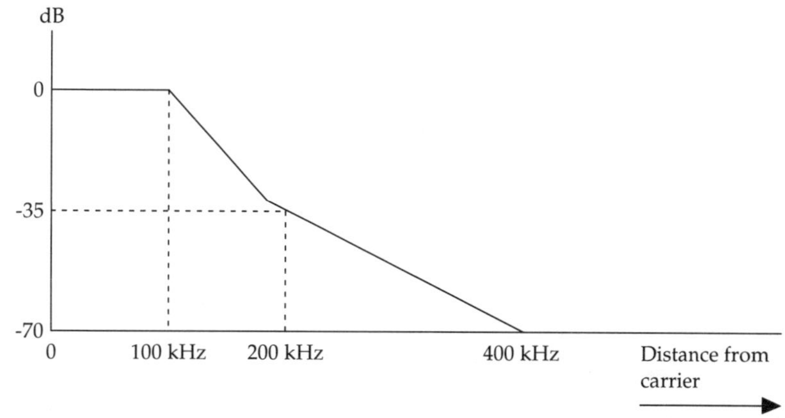

Figure 8.1

GSM is designed to use a 25-MHz up-link and a 25-MHz down-link with a base transmit from 935 to 960 MHz and mobile transmit from 890 to 915 MHz.

For countries with existing TACS networks there will be an extended migration period during which 10 MHz will be dedicated to the GSM down-link, 10 MHz to the up-link—the remainder will remain with TACS.

The 25 MHz is divided into 125 channels, each channel having a bandwidth of 200 kHz. Carriers 1 × 125 will not be used.

GSM uses GMSK modulation (see earlier chapter) modulated across the 200 kHz (see Figure 8.1) with some spectral spillage into adjacent channels.

MULTIPLE ACCESS STRUCTURE

GSM is a "signaling rich environment" with a multiple-access TDMA structure that allows plenty of conversation opportunity from base station to mobile and back again.

Each 200 kHz carrier is time divided into:

Multiframes	made up of 26 frames
Frames	made up of 8 time slots
Time Slots	made up of 8 fields

A time slot has 156.25 bits.

A frame has $8 \times 156.25 = 1250$ bits.

Bit rate is 270.833 kb/s tied to a 13 MHz clock $\frac{13\text{ MHz}}{48}$

Frame rate is 216.68 frames per second with every 13th or 26th frame dedicated to control.

Within a Time Slot

There are eight fields within a time slot. The 156.25 bits within the eight fields consist of 2×3 bit tail bits (TB) for synchronization, 1×8.25 bit guard period (GP) for power ramp up and down, 2×57 bit sets of coded data (representing the actual information—speech, text, or numerics—that we want to send), 1×26 bit training sequence for channel equalization and 2×1 bit stealing flags (SF) used when interleaving speech frames onto TDMA frames. (See Figures 8.2 and 8.3.)

It is necessary to "deliver" the time slot within a tightly specified power/time template maintaining level stability during transmission, timing within the time slot, and tightly constrained deviation depth, phase, and frequency errors. (See Figure 8.4.)

Within a Frame

There are eight time slots. The mobile is active in two time slots—one time slot for transmit, one time slot for receive. The mobile uses the six spare time slots to measure signal strength on five adjacent base stations and its own base station. (See Figure 8.5.)

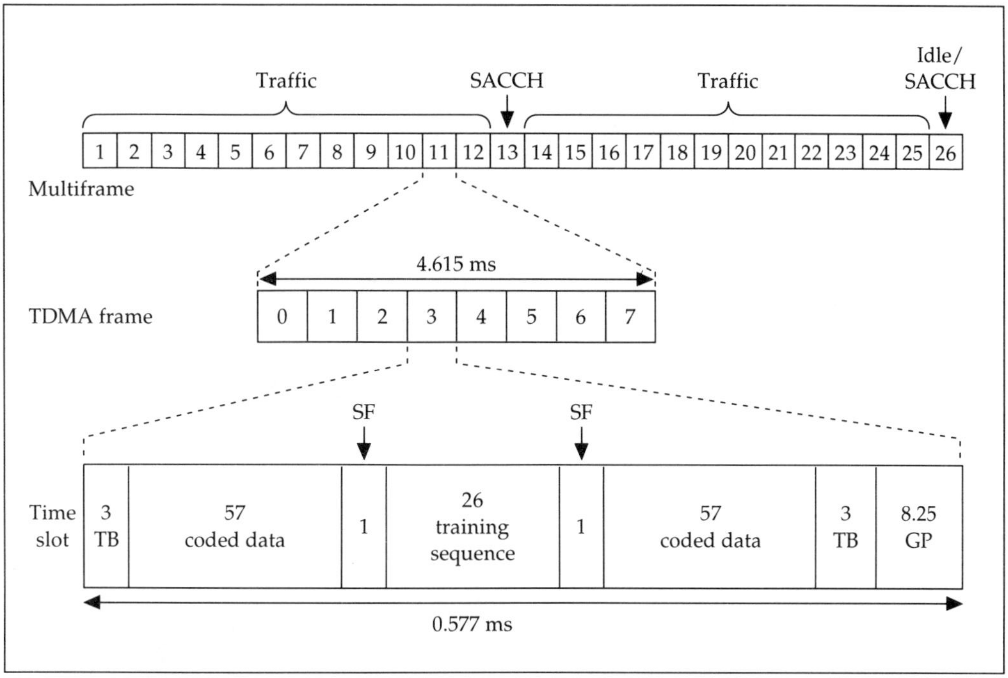

Figure 8.2

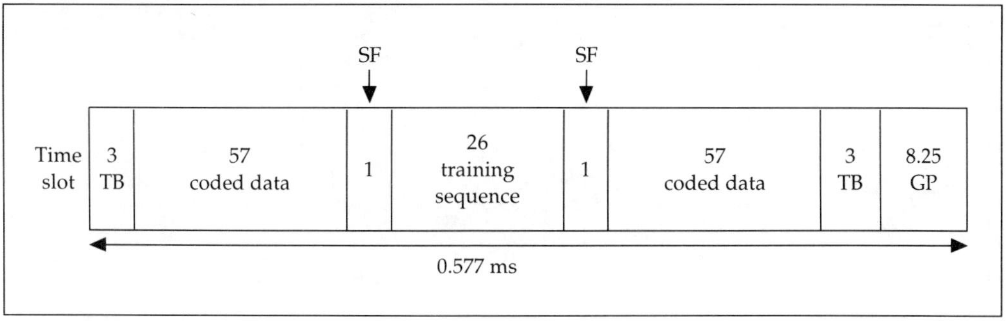

Figure 8.3

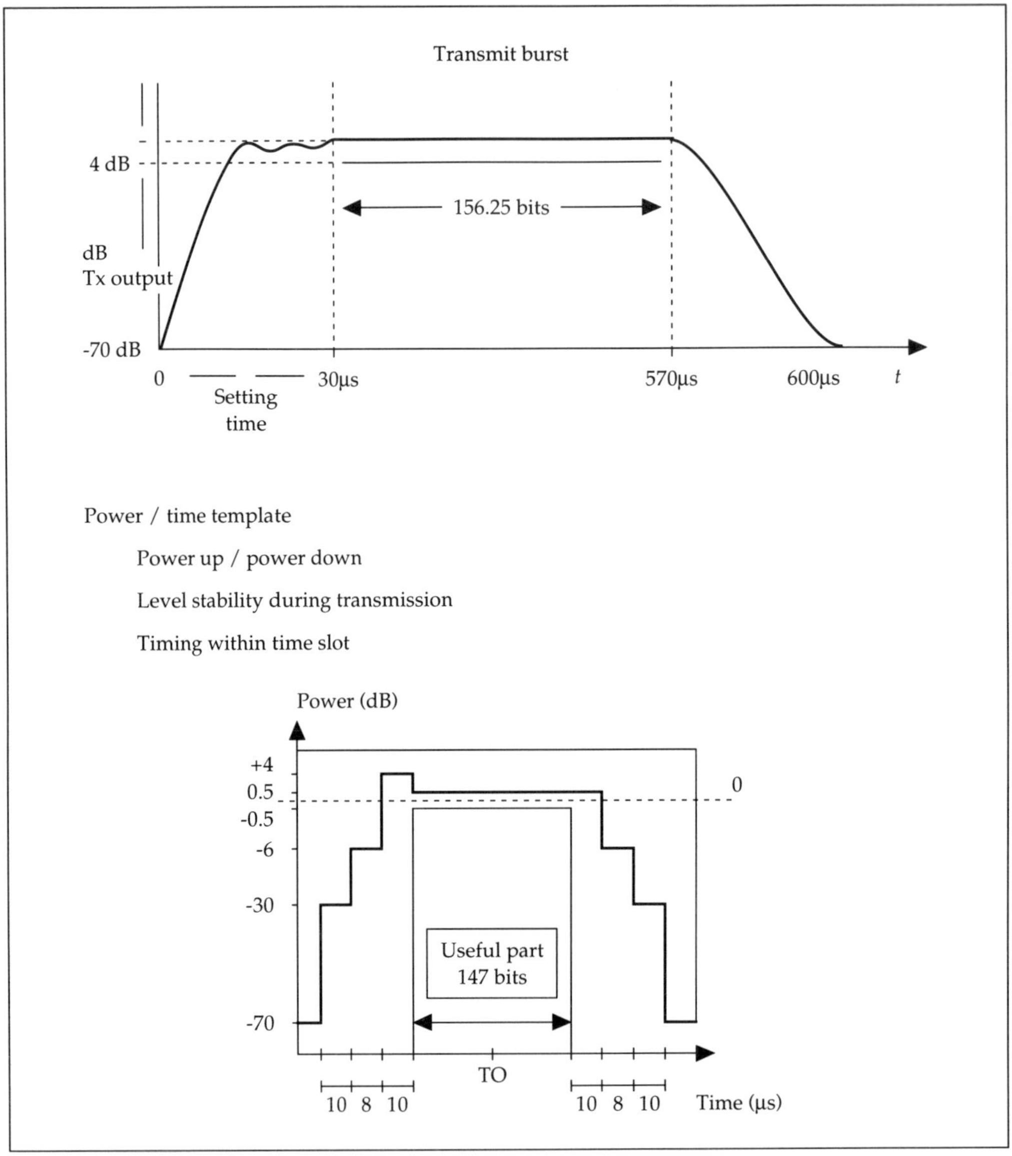

Figure 8.4 *"Delivering"' the time slot*

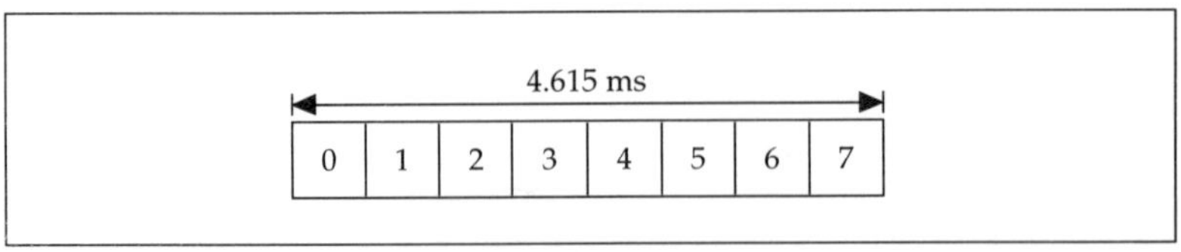

Figure 8.5

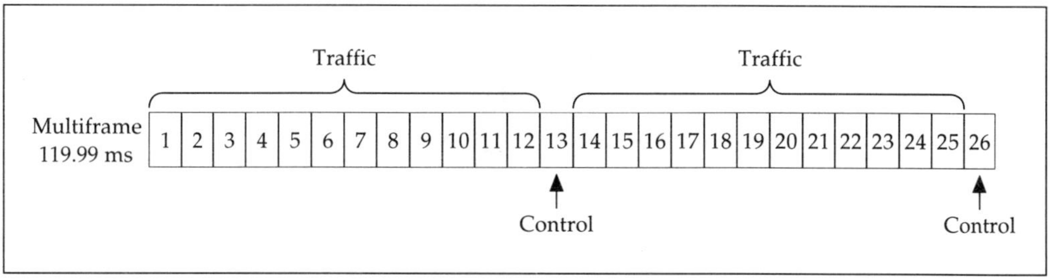

Figure 8.6

Within a Multiframe

Frames 1 to 12 are dedicated to traffic with eight traffic channels per frame. The 13th frame is used as a "slow-associated control channel" for the exchange of air interface/link quality information. The 26th frame is either idle or used for slow associated control—used by the mobile to identify adjacent base stations or file measurement reports to the base transceiver station (BTS). (See Figure 8.6.)

Power control and hand-off commands are down-linked via a "fast-associated control channel"—a type of blank and burst signaling (as in analog TACS) replacing a 456-bit speech block interleaved over eight frames.

Frequency Hopping

In severe multipath or conditions of severe co-channel interference, the mobile can be instructed to change frequency from frame to frame. Every mobile manufactured to the GSM specification must have this capability. The networks have the option of whether or not to implement the feature—frequency hopping will be difficult to implement in shared networks (e.g., TACS/GSM) with limited channel allocations.

The GSM Network

The GSM network consists of:

Mobile transceiver stations (MTS)
Base transceiver stations (one per cell) (BTS)
Base station controllers (BSC)
Mobile switch centers (MSC)

The MSC manages call set up, routing, termination, billing, and interface to the PSTN and data or private networks.

The Air Interface—Control Channels

Control channels are functionally defined as:

Broadcast (BACCH)	Performing frequency correction, synchronization and network information functions.
Common	Performing link establishment.
Dedicated (SACCH)	Performing slow-associated control.
(FACCH)	Performing fast-associated control.

NOTE The slow-associated control channel provides information back to the network for handover and power control decisions. The handover point by definition is triggered by low signal strength, low carrier-to-interference ratios and/or high BER (bit error rate) and/or adjacent/co-channel interference (most of these being interrelated).

The SAACH must therefore be robust and resilient and hence has a low data rate—383 bps net/950 bps gross—and substantial error control overhead.

Call Set Up and Call Maintenance

Call set up and call maintenance is performed as follows:

1. Find a Broadcast Channel
One rf channel (1 × 200 kHz) per site is dedicated to control. There are 32 predefined control channels in Europe. The mobile

scans control channel signal strength, synchronizes to the strongest channel, and receives a 51-frame multiframe from the BTS with frequency correction, synchronization, paging, access granting, and adjacent cell information.

2. Call Set Up
(Mobile initiated.) The mobile sends a message on a random-access channel (time slot 0 in a control frame). The BTS sends the mobile to an access grant channel (time slot 1 to 7 in a control frame). The mobile sends dialing information. The MSC manages call routing and finds the called party. A traffic channel is allocated (2 time slots in a traffic frame).

3. Call Maintenance/Clear Down
The slow-associated control channel every 13th frame monitors link quality. A measurement report is filed to the BTS. The BTS downloads power change or handover decisions on the FACCH. The network performs clear down and reallocation of the traffic time slot.

SPEECH CODING

Having defined (in summary) the multiple-access structure and signaling structure of GSM, we need to take a step-by-step walk through the various stages of digital processing from audio to rf and back again. First, the speech codec.

The object of a speech codec is to reduce the rate of the digital encoding process, deliver "toll"-quality speech—i.e., meet predefined subjective test mean opinion score criteria—produce an output that can be made relatively resilient to channel degradation, and produce speech quality that will degrade "gracefully" as bit error rate increases.

A human voice has tonal and temporal information content. Figure 8.7 shows the waveform map of a female voice saying "de" in the frequency and time domain.

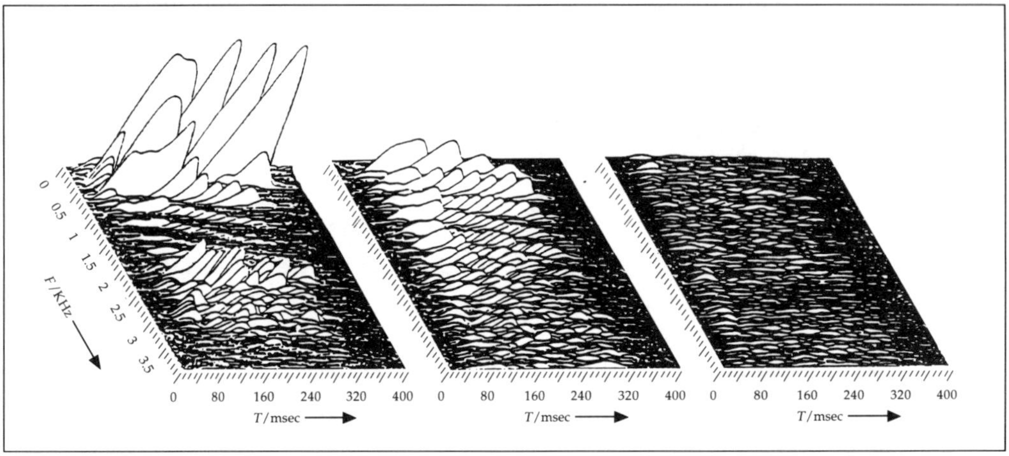

Figure 8.7

The waveform map in Figure 8.7 is a product of the vocal tract—a voicing source that produces vowel sounds and a noise source that produces consonants.

Codecs consider speech as being produced from a source filter—the vocal chords are a source, the vocal tract (throat, mouth, lips, tongue, teeth) a filter.

We said in the introduction that speech exhibits a high degree of predictability—i.e., speech is an *ergodic source*. Our codec strategy is to extract a low bit rate numerical expression for the tonal content, exploit the predictability of the source information from sample to sample, and find a numerically efficient method for expressing the periodicity, i.e., temporal information content.

THE GSM DIGITAL AUDIO INTERFACE

Our very first step is to take our analog speech waveform and sample the waveform at 8 kHz using a linear A–D converter (See Figure 8.8). The sine wave amplitude in each 125 μs block is quantized using a 13-bit code and gives us a bit rate output from our waveform coder of 13 × 8 kHz = 104 kb/s.

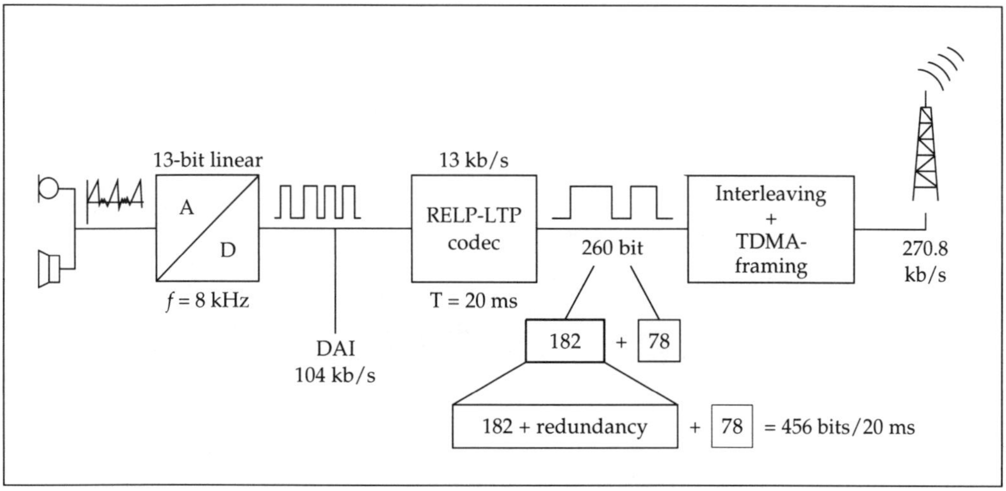

Figure 8.8

We input the 104 kb/s into our GSM speech codec, which analyzes the data stream using a combination of techniques known as "residually excitated linear prediction—long-term prediction" (RELP-LTP).

The 104 kb/s input bit stream is analyzed in 20 ms blocks. Taking an average within each block, the tonal information is described by eight digital filter coefficients.

The long-term prediction (LTP) process makes a correlation from sample to sample that is expressed as a correlation coefficient. The filter and correlation coefficients are transmited at 3.6 kb/s together with the residual excitation sequence (temporal information) at 9.4 kb/s giving the combined codec output rate of 13 kb/s.

CHANNEL CODING

To meet the resilience requirement for the speech codec each 20 ms block of 260 bits is divided into

182	most significant bits (class 1)
78	least significant bits (class 2)

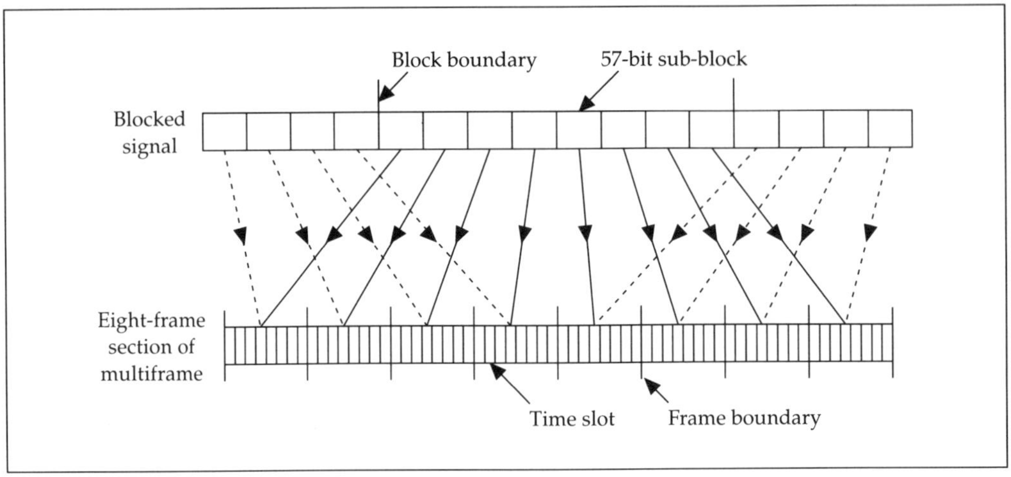

Figure 8.9

Errors in the most significant bits have a more noticeable impact on speech quality and therefore need additional protection. 192 forward error correction bits are added to the 182 most significant bits, then recombined with the 78 least significant bits to make up a 456-bit block.

Each 456-bit block is then split into 8 × 57 sub-bit blocks and interleaved over eight time slots and eight frames. (See Figures 8.9 and 8.10.)

THE GSM CHANNEL EQUALIZER

The GSM channel equalizer has to compensate for multipath delays equivalent to four bits. The 26-bit training sequence "nested" in the middle of a time slot provides information for estimating the transfer function in the time domain between the transmitter and the receiver. Once the transfer function is known, a prediction of received phase and amplitude can be made for each combination of bits that might have occurred; i.e., the equalizer transforms the received signal to the form it would have had if all the multipaths had been of equal temporal length.

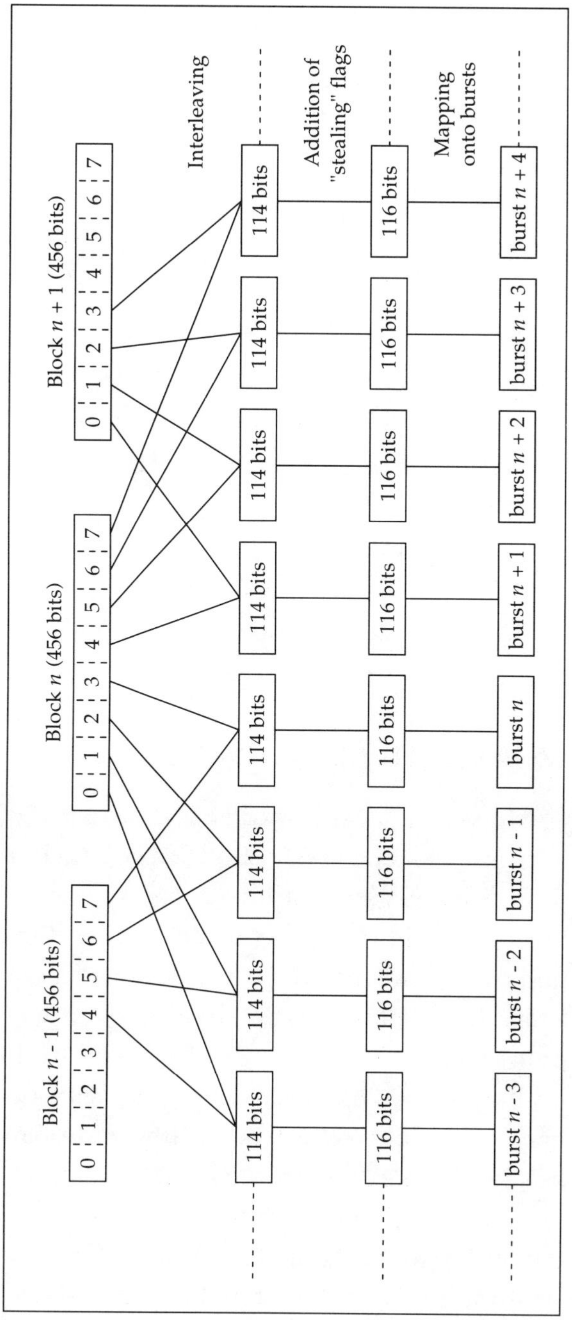

Figure 8.10 *Interleaving speech frames into TDMA frames*

GSM PROCESSING OVERHEAD

We stated that the output from the codec was 13 kb prior to channel coding. With eight speech carriers this would give us a net data rate at the air interface of 8×13 kb/s = 104 kb/s. The gross bit rate of the interface is 270.833 kb/s implying an overhead of 62% for error protection, control, and signaling functions; i.e., the gross rate per channel of $270.833 \div 8$ means a gross bit rate per speech carrier of 33.85 kb/s made up as follows:

Speech codec	13.0 kb/s		39%
		= 22.8 kb/s	
Codec error protection	9.8 kb/s		29%
SAACH	0.95 kb/s		2%
Guard time/ramp time/synchronization	10.1 kb/s		30%
	33.85 kb/s		100%

GSM BANDWIDTH GAIN

GSM uses a relatively wide band carrier—200 kHz as opposed to a traditional FDMA 30-kHz or 25-kHz carrier. This gives some advantage in multipath conditions in that only parts of the signal will be subject to destructive signal phase cancellation at any one instant. This is known as the *coherence* or *correlation bandwidth*—the bandwidth over which spectral components are affected in a similar way by frequency selective finding. For mobile radio this provides a definition of narrow band as $<$ coherence bandwidth, wide band as $>$ coherence bandwidth; i.e., with wide band only parts of the bandwidth will suffer fading at any one time as the receiver passes through the multipath environment. At 900 MHz, the correlation bandwidth is 100 kHz, therefore GSM by this definition is a wideband system.

NOTE It is sometimes asked why the frequency spread should make any difference, given that radio waves at any frequency travel at the same speed (at or near the speed of light). The effect is better understood by considering the relationship of frequency and phase. The speed of rotation of the phase vector is directly related to the frequency; therefore, the effect of signal phase cancellation becomes randomized across the frequency bandwidth.

GSM NETWORK OPTIMIZATION (MEASUREMENT REPORTS AND MAHO [MOBILE ASSISTED HAND-OFF])

In analog TACS or AMPS, neighboring base stations measure a signal transmission from a mobile and transfer the measurement information to the local switch for processing to make decisions on power control and handover. The information is then down-loaded to the mobile via the host base station. In an analog network being used close to capacity, this can result in a high signaling load on the links between the base stations and switches and a high processing load on the switch.

In GSM, the mobile uses the six spare time slots in a frame to measure the received signal strength on a broadcast control channel (BCCH) from its own and five surrounding base stations. The mobile then preprocesses the measurements by averaging them over a SACCH block making a "measurement report," which is then retransmitted to the BTS (base transceiver station) using an idle SACCH frame.

The mobile needs to identify co-channel interference and therefore has to synchronize and demodulate data on the BCCH to extract the base station identity code, which is then included in the measurement report. The mobile performs base station identification during the idle SACCH.

The measurement report includes an estimate of the bit error rate of the traffic channels using information from the training sequence/channel equalizer. The combined information provides the basis for an assessment of link quality degradation due to co-channel and time dispersion and allows the network to make fast, intelligent, responsive power control and handover decisions. The process is known as *mobile-assisted hand over*, or MAHO.

NOTE In rural areas there is the possibility that the network could decide on power control and handover by estimating the mobile-to-base-station distance and relative distance to other cells, calculated on the basis of the bit period delay of the received burst from the mobile on the reverse access control channel.

CAPACITY COMPARISONS

The design criteria for present analog cellular systems is to deliver acceptable speech quality in carrier-to-interference (C/I) conditions of 18 to 20 dB. The equivalent target for GSM without frequency hopping is 11 dB C/I or 9 dB C/I with frequency hopping making a three- or four-cell reuse plan feasible.

A capacity comparison, assuming 2% blocking and three sectors per site would be as shown in Table 8.1.

Table 8.1

	Analog FM		GSM	
		Pessimistic		Optimistic
Bandwidth	25 MHz		25 MHz	
Number of voice channels	833		1000	
Reuse plan	7	4		3
Channels per site	119	250		333
Erlang/km^2	11.9	27.7		40
Capacity gain	1.0	2.3		3.4

* **Source**: Raith & Udderfeldt. *IEEE Transactions on Vehicular Technology*, Vol 40, No. 2, May 1991.

The GSM comparison assumes a full-rate codec. A half-rate codec (11.2 kb/s gross rate) would increase capacity.

CONSTRAINTS TO CAPACITY GAIN

To maximize channel/spectrum utilization, digital cellular systems are designed to operate at low carrier-to-interference ratios and/or near the C/I margin—the difference between the C/I used for the frequency plan and the C/I at a predefined speech quality threshold. Working "close to the margin" demands fast, reliable power control and handover algorithms. The capacity limit is also influenced by the C/I rate of change; e.g., a vehicle-mounted mobile in motion in a dense urban environment results in a fast rate of C/I change.

The C/I target of 9 dB assumes the implementation of DTX—discontinuous transmission. DTX involves voice activity detection—no audio in and no rf output out in the time slot. This improves adjacent and co-channel performance (i.e., creates less interference to users) and reduces the power budget of the mobile, but DTX also reduces the confidence level of the signal level and link quality measurement reports. Early trials suggested some risk of a power lock-up condition in which all mobiles and all base stations moved to their highest power level (minimum attenuation).

The GSM Half-Rate Codec

As codec rate reduces, codec output becomes less resilient to channel degradation and bit errors. This can be offset by increasing the error protection overhead but this rather defeats the object of using a lower rate codec in the first place! Also, the lower the network's C/I ratio and C/I margin, the steeper the BER curves. It becomes difficult to implement a low-rate codec that delivers graceful degradation of speech quality.

Capacity Gain with Microcells

Mobile-assisted handover (MAHO) does assist the implementation of microcells. Street microcells are dimensioned by street corners with a typical 20 dB drop at a corner demanding a handover reaction time of one to two seconds. Co-channel interference conditions are very difficult to predict and demand a mix of conservative frequency planning at the microcell level and an overlay of umbrella cells. If a mobile is moving quickly from base station to base station (e.g., a car travelling fast in an urban environment), the mobile would be serviced by an umbrella cell. A stationary mobile or a pedestrian user would be serviced, if at all practical, by a microcell; therefore the power control/handover algorithm needs to be optimized to respond to the C/I "dynamic rate of change" and needs to be biased toward moving users down to microcell level out of the umbrella cells to increase channel reuse.

Adaptive Channel Allocation—ACA

Traditional frequency planning becomes quite difficult with microcells due to the unpredictability of co-channel interference. An option is to use *adaptive channel allocation* with no fixed frequency plan. Channels are allocated by assessing the real-time traffic situation and real-time interference in the cell, with the TDMA structure allowing the mobile and base station to search for better channels during a call. Advantages are minimal trunking efficiency loss and avoiding the need to design for "worst case" propagation conditions. Capacity gains of up to 50% are claimed for large systems with a three to four times capacity gain for small systems, and the technique is proposed for use in the Digital European Cordless Telephone system (DECT).

PERFORMANCE BENCHMARKS

Delay Budgets and Echo Cancellation

Transmission delay and echo, together, can make a significant impact on the perceived speech quality within a land transmission or mobile network.

Geostationary satellite links, for example, introduce a one-way path link and processing delay of 260 ms and are only barely tolerable. Without echo suppression, these would be unusable.

It will be quite possible that some GSM calls will be routed via satellite links (you can phone anywhere in the world from an analog or digital cell phone); i.e., GSM's own path and processing delay will be added to the "barely tolerable" satellite delay budget. Even in an all-terrestrial link, GSM's processing delay, added to the path link delay, will be significant.

Channel coding and TDMA frame interleaving in GSM introduce a one-way transmission delay of 37 ms. The speech codec has an algorithmic delay of 20 ms (the sampling slice) and an additional delay of 5 ms for processing. The digital links between mobile switch and base station, A–D and D–A conversions, equalization, channel encoding, and switching through the network add another 25 ms, adding up to a total one-way delay approaching 90 ms.

GSM, even on short path links, will need echo control nationally, but for calls routed internationally, an international gateway echo control device will be required. These were not part of the original GSM specification and are presently the subject of some discussion. One of the problems is that existing echo control devices have not been designed to work with non-linear processes such as low bit rate codecs. Echo paths may also change during a mobile call, so the echo control system must be dynamically responsive.

One answer would be to reduce GSM's processing delay—for instance by replacing interleaving with diversity reception and frequency hopping. Large interleaving depths (8 or 16 frames) are used in GSM to randomize error bursts in dynamic multipath conditions. Theoretically, interleaving could be replaced by frequency hopping or by a hybrid combination of low interleaving depth (two frames) and once-per-frame frequency hopping. Existing codec families could be replaced with faster codecs such as codebook excited codecs (as used in the United States ADC system).[1]

Performance Benchmark—Frequency Accuracy/Stability

The frequency accuracy and long-term stability requirements of the GSM network require the network to be locked at each base station either to a Caesium standard or comparable accuracy reference—for example the GPS (global satellite positioning system) standard.

Performance Benchmark—Power Budget/Size/Weight

GSM was originally planned to support car mobiles with a limited hand-portable user community. In practice, the rapid increase in performance of analog TACS, AMPS, and NMT (Nordic Mobile Telephone) hand-portables has unlocked a significant market demand for these products—for example, the Microtac Light at 220 gm with a 40-minute talk-time and 8-hour standby and Japanese equivalents. Portables in many markets—United Kingdom, Italy, Scandinavia, and many Asia-Pacific markets—are becoming predominant in the user community. There is therefore a clear market need for comparable GSM hand-portables.

In theory, GSM portables should be relatively power efficient, transmitting for only 1 slot in an 8-slot frame. In practice,

[1] For further information, see STS Chia, "Delay and Echo Control." *British Telecommunications Engineering*, Vol 10, July 1991.

the processing overhead cancels out much of this benefit. The GSM hand-held power budget can be reduced by implementing DTX (rf in the time slot only when audio input is detected) and DRX (discontinuous reception). Benefits from DRX depend on the number of paging channels but can result in a duty cycle ratio of 2% to 9%.

Major advances in reducing power consumption will depend on higher level circuit integration, power-saving device architecture and microprocessor control architecture and low voltage supply rails.

PERFORMANCE IMPLICATIONS

GSM places substantial demands on the ability of designers and the performance of devices.

Modulation

We discussed the deliberate spectral efficiency/power efficiency trade-offs implicit in the choice of Gaussian minimum shift keying (GMSK) as a modulation method and the problem of achieving an accurate frequency- and phase-referenced modulated carrier and (at the receiver) sufficiently accurate carrier recovery and demodulation techniques.

To summarize:

1. A phase-locked loop and traditional FM modulation create problems of accuracy, stability, and bandwidth with a data rate at the base station with components near dc causing problems.
2. Taking baseband mixed up to 900 MHz then IQ modulated risks imbalance in the mixer producing residual AM.
3. Using direct digital synthesis to address a sine wave look-up table provides phase coherence but cannot be realized at much above 100 MHz.

The answer (or one answer) is to combine the virtues of the direct digital synthesizer (DDS)—phase coherence referenced to a clock, with the virtues of the phase-locked loop/frequency synthesizer, i.e., high frequency/low noise and frequency accuracy referenced to a crystal to achieve an accurately phase- and frequency-referenced modulated carrier at 900 MHz—the process of direct digital interpolation (DDI).

Power Control

GSM demands not only frequency agility from the mobile and the base station, but dynamic power control. For example a base station could be transmitting on eight successive time slots at eight power levels. To implement efficient frequency hopping requires at least five carrier frequencies—one (fixed) for the control channel and four hopping channels. One proposal (Orbitel) is to switch at baseband then transfer to a row of fixed frequency rf carriers already settled on frequency. Note power control across channels also requires reasonable linearity in the PA chains.

Signal Processing

The GSM codec has been implemented on silicon by Analog Devices, AT&T, Motorola, Texas Instruments, and others.

Much work continues on codec optimization. Early experience proved that error correction is always needed to ensure the integrity of the data and that totally errored messages are better ignored than executed verbatim. (See GSM cyclic coding example in Chapter 5).

Early codec designs produced total silence in the speech gaps and were found to be uncomfortably unrealistic in practice, requiring the addition of "comfort noise."

Improvements could be realized by using floating-point as opposed to fixed-point arithmetic but although these codecs perform better in subjective tests, they perform less well than fixed-point codecs when exercized with the GSM-defined codec test algorithm.

The focus today is to reduce the size and power consumption of the DSP core and improve accessing and processing into and out of the high-speed RAM and ROM within the codec.

Channel Equalizer

The equalizer is one of the most complex single devices in the GSM phone—only realizable successfully by using the computationally complex Viterbi algorithm—computational complexity means cost and processing delay.

Software Engineering

Each mobile has to be capable of executing up to 40 MIPS (million instructions per second), responding to the "signaling-rich" hierarchy of GSM within very tight time constraints—hopping each TDMA frame every 5 ms, reporting radio power measurements every 480 ms and checking identity of surrounding cells every 30 seconds or so. Over 256 kB of on-board software implies complex processing—complex processing implies power consumption.

At the system level, GSM will need to be an intelligent network capable of capacity reconfiguration (concentration of capacity to meet sudden increases in demand such as natural or human disasters), with substantial network resilience (rerouting to avoid faults).

The Pan-European database will be one of the most complex databases ever established—for every 50,000 subscribers added to the network, the network will need to process 50 transactions per second. There could be 10 million GSM subscribers by 1999 across a minimum of 19 countries—billing and cross-billing at different rates and (probably) in different currencies.

FIRST-GENERATION GSM MOBILES

A GSM phone is an architecture divided into two parts: digital baseband and analog rf. As an example, Orbitel's 8 watt class 2 vehicle mounted and transportable GSM phone uses (for baseband) a dedicated ASIC for each air interface function; i.e., equalization, speech coding, channel coding, a dedicated timing chip for controling the microprocessor, an interface chip for the microprocessor and the 256 kB microprocessor itself.

Baseband ASICs use one μm (micron) and sub-micron technology. Analog devices are 1.5 μm. Second generation phones (1992) will use even smaller geometry technology and by 1993, the phone will be a two-part device—a complex analog chip for the rf section and a large microcontroller for baseband.

First generation Nokia GSM phones have a power budget of 100 mA on standby and 400 mA—4 A for a maximum of 8 watt rf output.

NOTE Nokia has spent 800 worker years to date on GSM development of which 200 worker years has been spent on their first generation phone. Technofone (now a subsidiary of Nokia) claims it takes 60 engineers at least three years to develop a GSM phone as opposed to 20 engineers taking two years to develop an analog phone.

GSM AS A COMPETITOR TO PCN

In the United Kingdom, Vodafone and Cellnet are intent on positioning GSM as a competitor to the PCN (personal communication networks) at 1.8 GHz with a GSM derivation called (by Vodafone) MCN (micro-cellular network)—using 800 mW class 5 mobiles for urban high-density applications. The idea is that GSM and MCN compete for bandwidth, and an MCN handheld up-grades to GSM when used in a car.

GSM AND SMART CARDS

Radiolinja in Finland announced Smart Cards for GSM ready for use in July 1991, using either a fixed internal smart card or external card as a subscriber identity module (SIM) with the subscriber number, home location register, level of service, and authentication information.

The concept is that a user can "enable" any phone in any country with his or her individual number and authentication/billing details, allowing for short-term hire at airports with the rented phone adopting the user's personal number.

THE GSM NETWORK

The GSM network has the following main components (see Figure 8.11):

AUC	Authentication center
BSC	Base station controller
BTS	Base transceiver station
DAI	Digital audio interface (104 kb/s)
EIR	Equipment identity register
GSM	Global system for mobile communication
HLR	Home location register
ISDN	Integrated services digital network
MS	Mobile station
OMC	Operating and maintenance center
PLMN	Public land mobile network (or private)
PSTN	Public switched telephone network
VLR	Visitor location register
TDMA	Time division multiple access
TCE	Transcoding equipment

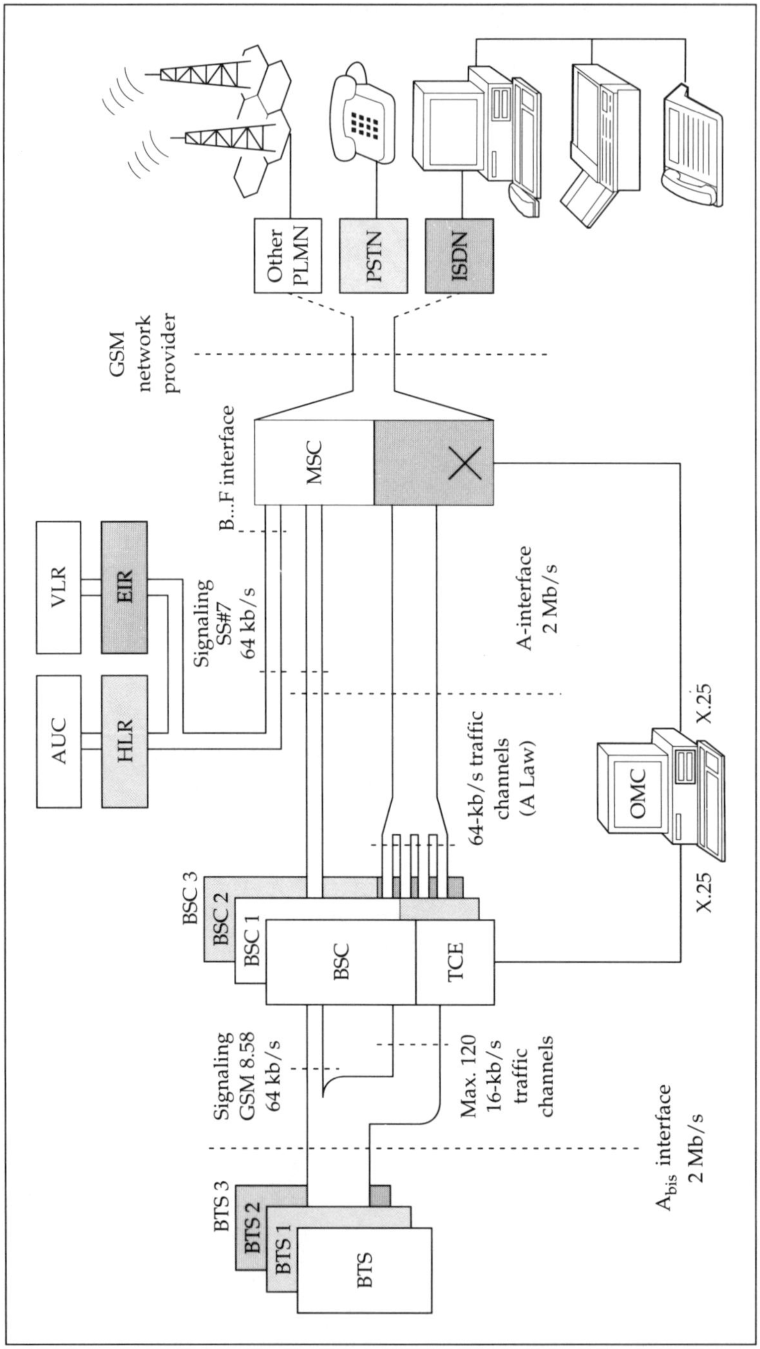
Other PLMN
PSTN
ISDN
GSM network provider
B...F interface
MSC
VLR
EIR
AUC
HLR
Signaling SS#7 64 kb/s
A-interface 2 Mb/s
64-kb/s traffic channels (A Law)
OMC
X.25
X.25
BSC 3
BSC 2
BSC 1
BSC
TCE
Signaling GSM 8.58 64 kb/s
Max. 120 16-kb/s traffic channels
A_{bis} interface 2 Mb/s
BTS 3
BTS 2
BTS 1
BTS

Figure 8.11

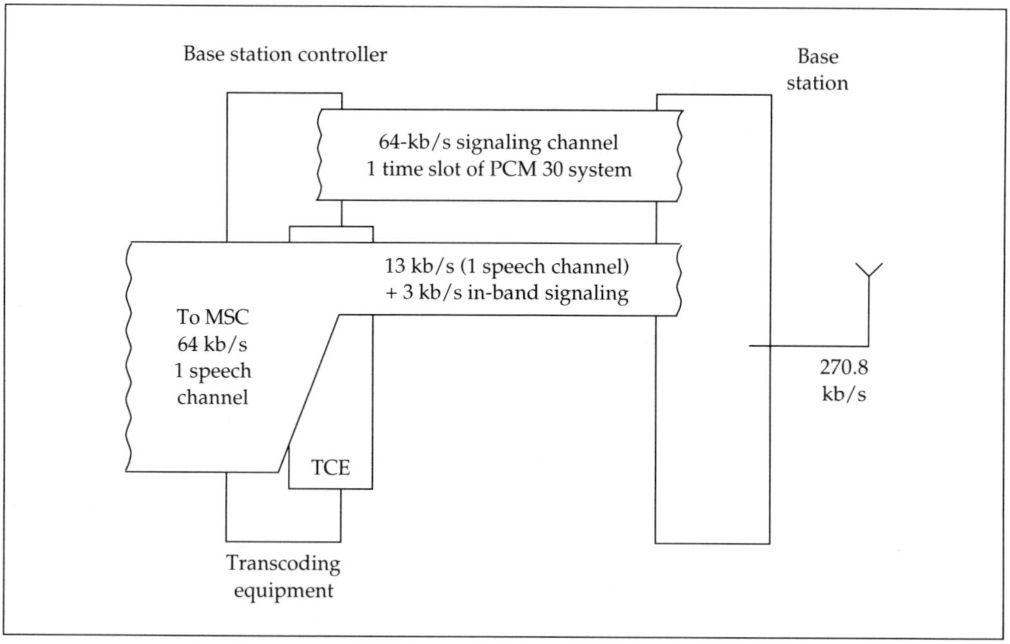

Figure 8.12

Base transceiver stations (BTS) and base station controllers use 2 mb/s CCITT G703 links consisting of 31 × 64 kb/s channel for speech transmission, each speech channel consisting of 13 kb/s plus 3 kb/s of in-band signaling giving four speech channels per 64 kb channel; i.e., 120 speech channels per PCM30 (30 × 4).

Transcoding equipment (TCE) translates 16 kb traffic/ speech channels to 64 kb/s a Law PCM. (See Figure 8.12.)

A 1 × 64 kb/s signaling channel manages call set up and clear down. A status report on receive level and link quality is forwarded by the BSC to the OMC (operations and maintenance center).

The signaling layer conforms to signaling System 7 (quite appropriate for mobile radio because of relatively long post dialing delay).

GSM AS A DATA BEARER

GSM data bearer services can be divided into:

> Transparent bearer services—where subscriber services are not resident within GSM.
>
> Nontransparent bearer services—where subscriber services are protected by a radio link protocol.

This is similar to analog TACS in the United Kingdom where Vodafone offers a nontransparent service (i.e., Vodafone adds the CDLC protocol or accepts input from a CDLC protocol modem and takes off the protocol prior to delivering to the PSTN or end user's PABX) or the transparent service of Cellnet where the user takes the responsibility of adding or subtracting mobile radio specific error coding at both ends of the link.

GSM's data bearer services are OSI (open standards interconnection) compatible in that users will be able to use a number of OSI Layer 7 applications such as X400 as well as ISDN supplementary services such as calling line identification (CLI). Synchronous and asynchronous data bearers are provided from 300 to 9600 bps (300, 600, 1200, 2400, 4800, 9600). Rates lower than 9.6 kb/s can use a full-rate or half-rate channel.

GSM data bearer services will interface with terminal equipment to CCITT V24/X21 with nontransparent services having the radio link protocol (RLP) added or removed at the translation gateways into and out of the network.

The RLP compresses four V110 frames into 200 bps with a 16-bit header, and 24-bit frame check sequence. The 240 bits are then interfaced to the radio system as four 60-bit blocks.

A half-rate convolutional code is applied taking the block length to 456 bits (four time slots with 114 traffic bits each). The frame is then diagonally interleaved over a number of TDMA frames.

GSM also maintains the option to deliver short status messages over the control/signaling channels (control channel time slot).

TESTING GSM

Two critical factors in GSM are to assess the resilience of receivers to bit error rate and frame erasure. A number of test strategies have been suggested.

Exercising of a Mobile by a Test Signal

The test signal is termed TU50 (typically urban 50 km/h) and is applied to a mobile at a level of –99 dBm at the nominal frequency, then down-stepped to the failure threshold.

Application of a Power/Time Template

Analyzes burst-building transmission time, power up and power down, level stability during transmission, and timing within the time slot.

Transmitter Phase Trajectory

A product of the bit pattern and modulation. With a test signal applied at –50 dBm, 294 phase samples are taken in each transmit burst, and a variance analysis is made on the frequency and phase errors.

Output Spectrum

Calculated by using phase and power values sampled over the transmit bursts. Within the signal, the encrypted data bits constitute the modulation spectrum; the rising/falling edges constitute the ramping spectrum.

Base Station Testing

In-service performance test parameters include phase error, frequency error, transmitted carrier power versus time, average transmitted carrier power, spectrum of modulation and transient signals, spurious sidebands and intermodulation products, and demodulation of burst signals. The wide dynamic range of BTS power outputs, over 100 dB, requires the signals to be prefiltered.

GENERAL PRINCIPLES WHEN TESTING GSM

Modulation Analysis

The objective is to compare actual amplitude and phase values with the ideal, computing the difference over a data burst and including peak and RMS values.

NOTE Modulation inaccuracies will disturb the ability of the equalizer to correct for multipath effects.

Most measuring systems translate the rf carrier down to an IF, at which point the signal can be sampled by an A–D converter and the samples stored for analysis (comparison with the ideal, to detect phase shift, carrier frequency error, and amplitude flatness).

Modulation Error

In a complex modulation signal there will be an actual signal, an ideal phasor vector and an error vector.

In DAMPS (American ADC standard), the error vector would need to be within ±12.5% of the ideal vector. In GSM (with constant amplitude GMSK), the error vector is resolved as phase error and amplitude inaccuracy.

Loop Testing

GSM radios should typically have a loop-back mode, in which the radio will recover, correct, and decode a test input signal even if the data does not make sense, then re-encode the data with parity and framing for return to the test system. Path distortion or additional test inputs can be applied without changing the test system.

Carrier Frequency

The instantaneous frequency deviation due to phase modulation is often greater than the allowable carrier frequency error.

The difference between the actual and ideal phase trajectory produces a straight trend line, the slope of which is proportional to the carrier frequency error.

Spurious Emissions

These can be generated by local oscillator impurity, modulation products, and/or transmitter power on/off transients, of which the latter is often dominant but present for only a fraction of the time.

Transients may therefore be measured separately using a spectrum analyzer with a sweep rate sufficient to ensure a peak reading across the band (and avoiding aliasing effects where the sweep rate is an exact multiple of the burst rate).

Measuring the modulation products on their own involves the removal of the transients, achieved by use of a "soft gate" synchronized to the burst and open only during periods of modulation.

Problems with Mobiles

The equalizer will be upset by uneven burst flatness—caused by semiconductor heating or poor power supply regulation. Switching transients, like nonlinearity, may cause adjacent channel interference.

Faults in A–D converters, equalizers and data filters will all contribute to eye closure and hence to insensitivity or receiver "deafness." Lock detection thresholds in weak signal conditions are critical. The D–A converters themselves may become nonlinear. Mixers (with hardware modulators) may need balancing.

The GSM specification provides for an "electronic man–machine interface," EMMI, which allows a test system to press handset keys remotely, which is good for rapid type testing.

For in-service tests, manufacturers and network operators may prefer many of the tests to be carried out "off-air," i.e., the network monitors the performance parameters of mobiles registering onto the system and using the system.

In analog TACS and AMPS, network operators have found that major improvements can be made in overall network performance if the mobile user population can be monitored, measured, and barred from the network if the mobile(s) drift outside the network's defined performance parameters. For instance old analog AMPS phones in Chicago were discovered to be up to 8 kHz off nominal frequency—a potential threat to network integrity.

Maverick GSM mobiles may pose a similar threat. No one knows for sure what the long term frequency and phase stability characteristics of GSM phones will be through life. The network operators would prefer to keep control of the performance parameters of their mobile user communities.

In the United States, there is the potential for "economic harm litigation" if manufacturers deliver product to the networks that fails to meet long-term performance/reliability targets.

Two-chip mobile phones (third-generation GSM) may in any case, be economically and practically unrepairable at the dealer level, and the manufacturers themselves will probably prefer to collect and process any faulty phones in order to track any potential long-term performance problems.

GSM AND AUTOMOTIVE INTEGRATION

Car manufacturers worldwide, including Rover Group, UK and Ford of Europe, are beginning to take a keen interest in the mobile phone. Most executive cars can now be purchased with a factory-fitted cell phone and/or fax, with some manufacturers (e.g., Ford, UK) offering air time. Benefits to users include better mechanical and electrical integration into the car, factory-fitted feeder and antenna installation, and acoustic modeling of car interiors to improve hands-free operation. Factory-fitted cell phones also minimize the problem of acoustic coupling between mouthpiece and earpiece in conditions of relatively high ambient noise.

NOTE Sound travels at approximately 331 meters/second or 1 meter in 3 ms under normal conditions. The echo path due to the in-car acoustic environment can add an aggregate transmission delay of 10 to 20 ms to the overall delay budget.

The next generation of world car will be disposing with the traditional wiring loom—heavy, bulky and expensive (there are 2 to 3 kg of copper in an average American car). Lights, engine management, and all electrical subsystems will be controlled from a bus backbone, and a GSM phone will be part of the integration process, usable out of the car as, for example, a class 4 or class 5 hand-portable or in-car as a class 2 mobile, validated for use by the automotive smart card, matching the owner/user to the vehicle to the phone.

Because GSM is also a relatively efficient data bearer, it provides the opportunity for two-way telemetry links with engine management and monitoring systems, already widely used in Formula 1 racing. Mobile car servicing could take on a whole new meaning.

Note also that companies such as Ford are already substantial manufacturers of FM radio products (Ford is the world's largest manufacturer of FM radios). It has the technology needed to build a cell phone and/or integrate the phone with the FM radio for use in the car or as a portable radio phone/cell phone music-while-you-talk–while-you-walk device.)

In the United States, Ford and General Motors (with their E-TDMA proposals) are working toward similar levels of integration and added value facility.

GSM—EUROPEAN PARTNERS AND FISCAL IMPLICATIONS

GSM has, and always will have, a political dimension. Cellnet and Vodafone are soft-pedaling their introduction plans until 1993/94 or to immediately pre-date any PCN competitive roll-out. Cellnet will use Motorola as a main contractor with Nokia as a second supplier.

Motorola is cross-licensing with Matra and Alcatel on GSM/PCN patents.

In France, the GSM competition to Radiocom 2000 (SFR) will be provided by Alcatel. In Germany, the GSM network competing with Deutsche Bundespost, the D2 Privat owned by Mannesmann/Mobilfunk, will use Ericsson technology. Other players include Northern Telecom and its subsidiary BNR (Bell Northern Research) and Fujitsu working with Ericsson, Alcatel, and Siemens. GSM technology, as exemplified by Germany's D2 Privat network, is seen as offering some competitive edge over existing analog systems—not hard in Germany where the analog system is, or has been, severely over-subscribed. We see the beginnings of much closer Transatlantic links between Europe and the United States, with a visible growth in activity from, for example AT&T, Northern Telecom and Pactel (Pacific Telesis); e.g., Pactel's involvement in D2 Privat.

GSM's technology advantage will not necessarily be that clear-cut in countries with existing analog cellular networks adequately engineered to support the existing mobile user community. The two analog United Kingdom networks with 1320 25-kHz channels allocated are very capable of supporting up to 1.5 million users, delivering very acceptable speech quality and very acceptable service levels (acceptably low rate of blocked calls, dropped calls or lost calls). Note that user quality perception in analog cellular networks in the United Kingdom and United States is improving on a monthly basis, reducing the market imperative for implementing GSM, or equivalent digital cellular options.

The concept of GSM is that it can use the same base stations as the existing analog TACS network but support two to three times as many users per base station. Provided two to three times as many users can be sold onto the network, GSM will make financial as well as technical sense.

It is appropriate at this point to compare GSM's technology platform and "product offering" with the RoW (rest-of-the-world) digital cellular options.

CHAPTER

9

DIGITAL CELLULAR COMPARISON

To an extent, digital cellular is the product of the standards making industry; GSM is the product of ETSI (European Telecommunications Standards Institute); ADC (American Digital Cellular) is the product of the TIA (Telecommunications Industry Association); JDC (Japan Digital Cellular) is the product of the Japanese Ministry of Post and Telegraph (MPT).

ADC

American Digital Cellular is a dual-mode network retaining analog control channels. Single-mode analog phones access the 30-kHz analog traffic channels via the analog control channels. Dual-mode digital mobiles access via the analog control channels but then use analog and/or digital traffic channels.

There are three (full-rate) or six (half-rate) TDMA digital traffic channels available for every 30 kHz of bandwidth allowing for a 3-for-1 or 6-for-1 rack replacement on a channel-by-channel basis at the base station; i.e., analog traffic channels are gradually replaced by digital traffic channels offering a network-managed as opposed to market-managed migration.

NOTE

1. In the longer term, the analog control channels will need to "go digital" to improve access efficiency and to reduce the cost, weight, and size of the mobiles, i.e., single-mode digital mobiles.
2. The United States market has traditionally been based on a smaller percentage of hand-portables and a larger percentage of car phones, so weight and size have been less of an issue than in European markets. Americans (as a general rule) are less pedestrian than Europeans; i.e., Europeans "walk and talk" more than Americans. Some Asian-Pacific markets, e.g., Hong Kong, are almost totally dedicated to hand-held phones—in-car phones are a rarity.

Modulation Method in ADC

ADC opts to use π4 DQPSK, a four-level modulation method against GSM's GMSK two-level modulation, trading a higher gross bit rate against some intersymbol interference overhead

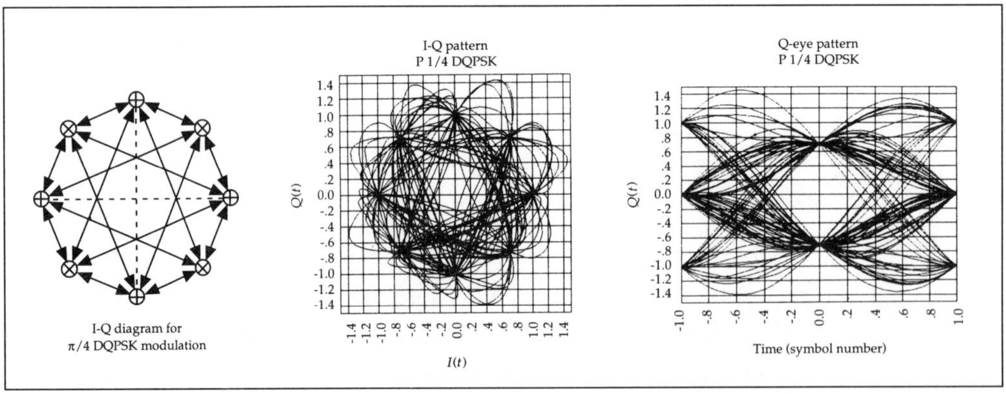

Figure 9.1

and envelope variation (needing more linear, less power-efficient PA chains). The overall trade-off gives some performance benefit over GMSK.

JDC—JAPAN DIGITAL CELLULAR

JDC will be similar to ADC but not dual mode; i.e., there will be no backward compatibility. JDC will use the existing FDMA 25-kHz spacing or 2 × 25 kHz FDMA channels interleaved. The bandwidth will be divided into three or six voice channels or 12 voice channels (50 kHz). JDC will be a politically managed migration, as opposed to network-managed migration (ADC), as opposed to market managed migration (GSM).

Air interface characteristics of the three digital standards are shown in Table 9.1.

Table 9.1

	Europe (ESTI)	**North America (TIA)**	**Japan (MPT)**
Access method	TDMA	TDMA	TDMA
Carrier spacing	200 kHz	30 kHz	25 kHz
Users per carrier	8 (16)	3 (6)	3 (tbd)
Modulation	GMSK	π/4 DQPSK	π/4 DQPSK
Voice codec	RPE 13 kb/s	vselp 8 kb/s	tbd
Voice frame	20 ms	20 ms*	20 ms
Channel code	convolutional	convolutional	convolutional
Coded bit rate	22.8 kb/s	13 kb/s	11.2 kb/s
TDMA Frame duration	4.6 ms	20 ms*	20 ms*
Interleaving	=40 ms	27 ms	27ms*
Associated control channel	extra slot	in slot	in slot*
Handoff method	MAHO	MAHO	MAHO*

* Ericsson Proposal

NOTE

1. GSM, ADC, and JDC all retain duplex spacing, 45 MHz for GSM and ADC and 130 MHz for JDC with new frequencies allocated above and below the existing cellular bands, so duplex filters (expensive and power hungry) will still be needed through all three networks.
2. Strictly defined, GSM, ADC and JDC are all hybrid TDMA/FDMA networks, retaining frequency division of carriers on the up-link and down-link.
3. Japan has allocated spectrum at 1.5 GHz for JDC (in addition to 900 MHz allocations) and 1.9 GHz for digital cordless.

ADC AIR INTERFACE

ADC has a slot and frame structure for the down-link and up-link with 324 bits per time slot (6.67 ms) and three or six time slots per frame depending on whether the network is full rate (three slots) or half rate (six slots). (See Figure 9.2.)

Each burst of 6.67 ms fits within a frame length of 20 ms. As a rule of thumb, 25 MHz should support up to 2500 traffic channels.

A gross bit rate of 48.6 kb/s is divided as follows:

Speech codec	7.95 kb/s		49%
Error protection	5.05 kb/s		
SAACH (gross rate)	0.60 kb/s	Overhead	51%
Guard time/ramp/ synchronization	2.60 kb/s		
Total per channel =	16.2 kb/s (gross)		
Total bit rate =	48.6 kb/s (3 × 16.2)		

i.e., ADC has a 51% overhead compared to GSM's 62%.

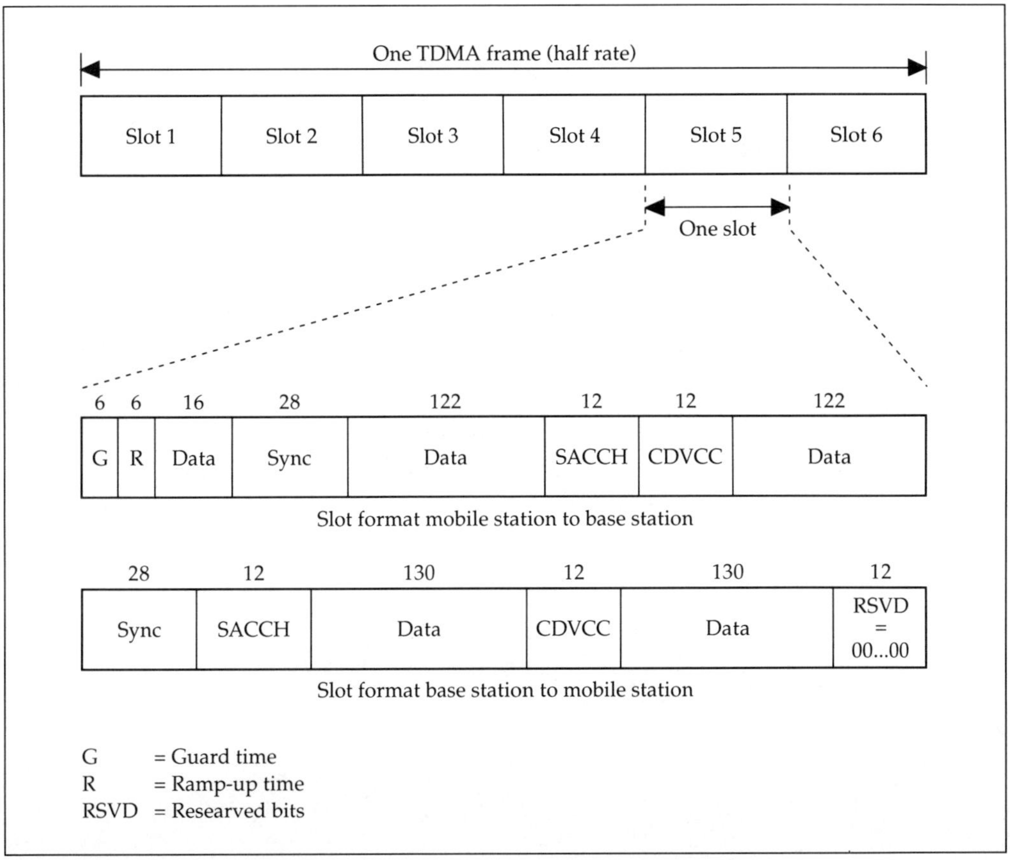

Figure 9.2

CO-CHANNEL IDENTIFICATION

Co-channel identification is performed within a burst using an 8-bit identifying color code protected by a 4-bit Hamming code making up a 12-bit word (CDVCC).

CHANNEL CODING

The speech codec processes 20 ms speech blocks, outputting 159 bits divided into class 1 (most-significant) and class 2 (least-significant) bits, which are then convolutionally coded and diagonally interleaved over 2 slots.

The SAACH bits are interleaved over 22 slots at a net rate of 227 bps.

ADC MAHO—(MOBILE ASSISTED HAND-OFF)

As there are no digital control channels, there is no extraction of a base station identity code. There are no idle frames and neighboring base stations identify a specific mobile by using a unique synchronization word to identify the time slot and the CDVCC to identify an intended user from a co-channel—i.e., matching the user to the channel.

Measurements from all base stations—up to 12 channels—are reported and compared. Measurement reports will tend to be more accurate in GSM, resulting in faster, more accurate power control and handover than in ADC but the difference will probably be marginal.

ADC VERSUS JDC VERSUS GSM

ADC will implement either a Motorola 8 kb/s or Ericsson 8.7 kb/s codec using deterministic code-excited linear prediction (a repertoire of waveform values stored in a waveform map/look-up table and digitally addressed). The ADC codec is convolutionally coded to give a gross rate of 13 kb/s compared to GSM's 22.4 kb/s.

NOTE The lower bit rate codec will tend to be less resilient to channel degradation. Interleaving in GSM is also more powerful (greater interleaving depth) than in ADC, giving more protection against fast, severe multipath fading. This will be offset to some extent by ADC and JDC implementation of antenna diversity giving approximately three to four dB of C/I gain. GSM can implement a half-rate codec; ADC/JDC can implement a half-rate codec.

ADC will implement channel equalization using the in slot "sync" sequence as a reference. JDC will most likely not implement equalization but will use diversity reception with a dual receiver chain. Note that "equalization" is less effective with narrow-band ADC/JDC compared to wide-band (200 kHz) ESM.

ADC/JDC will probably be able to implement a 4-cell frequency reuse plan.

The capacity comparison in Table 9.2 is suggested. Design criteria for current analog cellular systems 18 dB C/I over 90% of cell area are given in Table 9.3.

Table 9.2

GSM		ADC		JDC	
With frequency hopping	Without frequency hopping	With antenna diversity	Without antenna diversity	With antenna diversity	Without antenna diversity
9 dB	11 dB	12 dB	16 dB	13 dB	17 dB

Table 9.3

	Analog	**GSM**		**ADC**		**JDC**	
	FM	(1)	(2)	(1)	(2)	(1)	(2)
Bandwidth	25 MHz	25 MHz		25 MHz		25 MHz	
Number of voice channels	833	1000		2500		3000	
Reuse plan	7	4	3	7	4	7	4
Channels per site	119	250	333	357	625	429	750
Erlang/km^2	11.9	27.7	40.0	41.1	74.8	50.0	90.8
Capacity gain	1.0	2.3	3.4	3.5	6.3	4.2	7.6

* 1. Pessimistic
2. Optimistic

Assumptions—cell radius of 1 km (site distance 3 km), 2% blocking.
Source: Raith & Uddenfeldt (Ericsson) *IEEE Transactions of Vehicular Technology* Vol. 40, No. 2, May 1991.

Digital speech interpolation (DSI) increases channel capacity by taking advantage of the quiet times that occur in speech.

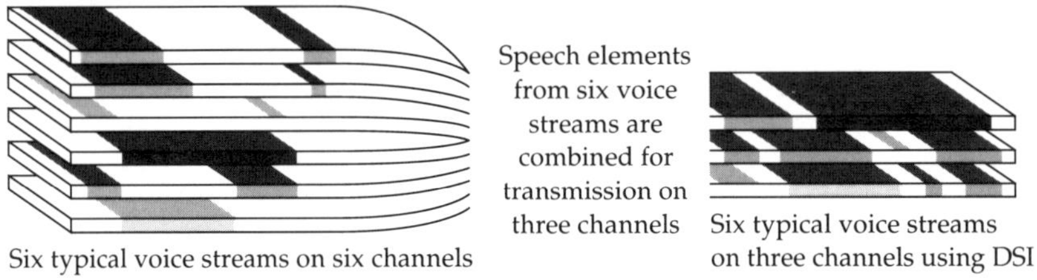

Figure 9.3

OTHER UNITED STATES TDMA OPTIONS

With many U.S.A. cellular networks financed by high-interest loan stock, stockholders have a particular interest in technologies that promise a higher rate of return on a given amount of spectrum real estate (spectrum allocation). The technology's promise to deliver can be used as loan or funding collateral. This can lead to a tendency to over-sell the capability of competing technologies, but (looking at the positive side) it does encourage the development of very spectrum-efficient technical solutions.

E-TDMA, promoted by Hughes and General Motors in the United States, is an example of just such a "leading edge" technology; taking the TDMA concept of Ericsson but implementing half-rate coding and digital speech interpolation (reallocation of time slots when no speech input is present). A 15-times capacity increase is claimed. (See Figure 9.3.)

DIGITALLY ENCHANCED ANALOG OPTIONS

A proprietary route to capacity gain promoted by Motorola and implemented as N-AMPS (in the United States) and N-TACS (in Japan) divides 30 kHz analog traffic channels (United States) into 3 × 10 kHz analog traffic channels or 25 kHz analog traffic channels (Japan) into 2 × 12½ kHz analog traffic channels. Seven sub-audio digital bit streams and a digital ST stream replace the three traditional SAT tones and ST tone of analog cellular TACS and AMPS systems to extend the co-channel (co-SAT) distance. (See also DCS example in Chapter 6).

CDMA is a further option for capacity gain, meriting special attention (see Chapter 10).

For the moment, we will leave digital cellular to evaluate cordless technology options.

CHAPTER

10

CORDLESS TECHNOLOGY

Europe is in the grip of two competing cordless technologies, CT2 (cordless telephone 2) and DECT (Digital European Cordless Telephone). CT2 has suffered in the United Kingdom with lack of market acceptance and the (related) problem of system incompatibility. We argue that CT2 should not be dismissed as a failure but should be reevaluated as an "appropriate technology" with considerable application potential.

CT2

Much of the processing overhead in digital cellular is the result of the wide application specification—large cells, small cells, tracking mobiles migrating from cell to cell, power control, and hand-off algorithms. CT2 has, by comparison, a network hierarchy with no mobility management overlay (no outgoing calls to the mobile so no need for the network to track the mobile), short path link lengths (up to 200 meters) and a more-or-less stationary user (i.e., no dynamic multipath to correct).

Short path links mean a lower power budget allowing for compact, light-weight portables with extended (relative to digital cellular) talk and standby time, typically a 3-hour talk-time on rechargeables or 10-hour talk-time on primary batteries in a package weighing less than 200 gm.

The idea is that a domestic or office cordless CT2 will either access or take calls from a dedicated local base station (in the house, in the office) but can also make outgoing calls at a phone point—station concourse, cafe, or road-side site.

CT2 has a simple rf architecture—40 × 100 kHz channels between 864 and 868 MHz, using Gaussian filtered binary frequency shift keying with bit transitions constrained to be phase continuous.

CT2 has a simple digital processing architecture—the input speech waveform is waveform coded (quantized) using an adaptive differential pulse code modulation (ADPCM) coder with a bit rate of 32 kb/s. The two-way full-duplex conversation is then processed using time division duplexing (TDD). The 32 kb/s digitized speech signal is transmited at 64 kb/s. Each 2 ms of speech is transmited in 1 ms with the "gap" used for the speech return path—no need for "paired" frequencies or a duplex filter. (See Figure 10.1.)

The speech channel rate of 64 kb/s has 8 kb/s of signaling added to it to gross up to a combined rate of 72 kb/s; i.e., channel coding and signaling overhead is only 12% of the total bit rate.

The delay budget is 5 ms, 1 ms for time division duplexing, 4 ms for the ADPCM coder.

CT2 can be realized with a very simple processing architecture using a Homodyne receiver. 32 kb ADPCM chips are commonly available, robust and low cost—the output from the

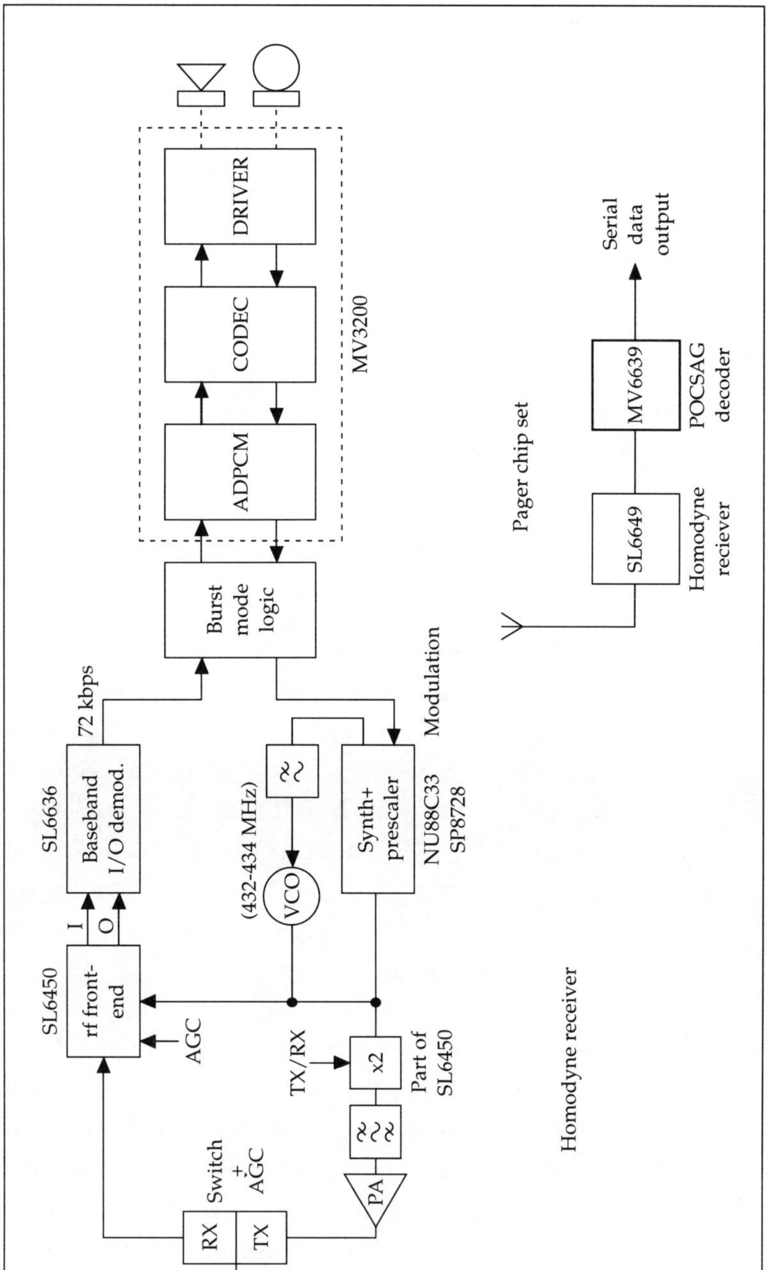

Figure 10.1 *Digital cordless telephone—UKCT2; Simple block diagram—generation 2*

coder is processed through some (relatively simple) burst mode logic and then IQ modulated onto the carrier.

The simple internal architecture of a CT2 phone lends itself to simple, low-cost volume manufacturing and is of particular appeal to manufacturers, often Asian-Pacific based, wanting to diversify away from analog cordless (CT1) or paging manufacturing. Homodyne techniques (with no intermediate frequency stage) are already well understood and easily implemented by paging manufacturers.

CT2 phones are therefore a product that can potentially be produced in very large numbers at a very low cost. If this happens, or is allowed to happen, CT2 will enjoy a renaissance. Domestic or office users would buy a CT2 phone in preference to a CT1 analog phone in very large numbers if the price point were the same or similar—there are 10 million CT1 phone users in Europe with 1 million sales per year so the market is already well established. The cordless penetration (users per 1000 of population) in the United States is substantially higher, with cordless sales projected to overtake corded sales in the early to mid-1990s. Domestic or office users owning a CT2 phone, which is battery powered, light weight, and compact with a longer talk-time than an equivalent cellular phone, could presumably be persuaded to use the phone for making outgoing calls at a public CT2 phone point and may be prepared to pay a small premium for the service.

CT2 has also been used with some success to provide cordless local area speech networks, interfaced to a PABX in offshore and onshore closed user group applications.

CT2 is not, and probably never will be, a system for delivering data. Data could, in theory, be delivered using a modem or using the 32 kb digital voice channel directly but this would not be ideal. As a speech bearer, however, CT2 still has much to offer as an "appropriate to task" technical solution.

Table 10.1

		DECT	CT2
Speed		1.152 mb/s	72 kb/s
Power			
	Peak	250 mW	10 mW
	Mean	10 mW	5 mW

DECT

DECT, in comparison, is much more comprehensively specified—a product of the ETSI standardization process (the ETSI—RES3 Radio Equipment and Systems Committee). Comprehensive specification means processing complexity and cost.

DECT combines FDMA with TDMA with TDD, taking 20 MHz (1880 to 1900 MHz) divided down into 12 radio carriers, each carrier supporting 12 time division multiplexed channels processing a speech data rate of 32 kb/s within a total channel rate of 1.152 mb/s.

DECT is designed as a speech bearer and as a data bearer. Time slots can be concatenated to increase bandwidth for high data rate transfer applications. (See Figure 10.2.)

The comparison in Table 10.1 also shows the wider dynamic power range of DECT compared with CT2.

DECT uses GMSK modulation and has a one-way delay budget of 12 ms. It is designed to be applicable for indoor pico calls and outdoor microcells, using adaptive channel allocation (ACA), with power control and handover capability.

CT3 is an intermediate protocol, proprietary to Ericsson, promoted as an evolutionary route to the full specification DECT systems.

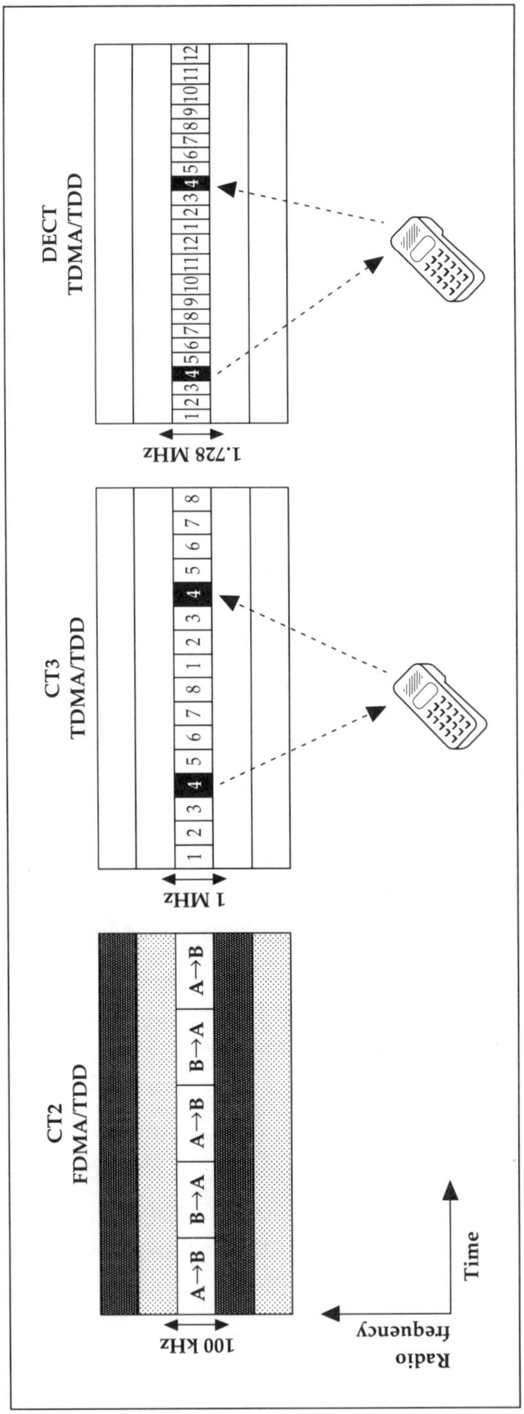
CT2
FDMA/TDD
A→B
B→A
A→B
B→A
A→B
100 kHz
CT3
TDMA/TDD
1 MHz
DECT
TDMA/TDD
1.728 MHz
Radio
frequency
Time

Figure 10.2

CORDLESS LOCAL AREA NETWORKS

Cabled local area networks (LANs) have developed since the 1970s as a delivery mechanism for data transfer between computers and computer peripherals. Bit rates are typically up to 1 mb/s for electronic mail (E-mail) or printer sharing, up to 10 m/bit/s for file transfer and 20 mb/s and above for video/ image transfer or "backbone" applications.

Commercial cabled LANs include Apple Talk for low bit rate applications (under 1 mb/s), Ethernet and Token Ring (up to 20 mb/s) and fiber optic for high-speed data transfer (up to 100 mb/s and beyond).

"Wireless" or "cordless" LANs (CLANs) offer the opportunity to relocate computers and computer peripherals within the office environment without expensive recabling/rerouting.

"Wireless" or "cordless" LANs must be capable of adapting to the changing propagation conditions within an office caused by changes in configuration (relocation of partitions for example), or dynamic changes caused by (for example) people moving around, a requirement that has prompted several manufacturers of CLANs to use spread spectrum code division multiple-access techniques.

Spread spectrum is addressed in more detail in the next chapter but essentially involves the transmitted signal being spread over a bandwidth much wider than the information bandwidth, avoiding the frequency/phase specific effects of multipath signal phase cancellation.

Spread spectrum also allows these low-powered networks (milliwatts rather than watts) to coexist with other narrow band users sharing the same allocated frequencies.

In Europe, it has been anticipated that such systems could work within the ISM (industrial, scientific, and medical) spectrum allocation at 2.45, 5.755, or 24.125 GHz (now implemented in the United States).

The products are typically packet based (individually addressed packets of varying size) and are compatible with most of the proprietary LAN open system software interfaces—

examples include the IT Security Radio Link, NCR Wave Lan product, and Telesystems ARLAN (Advanced Radio LAN). Radio Link delivers a data rate of 250 kb/s, Wave Lan 1 mb/s and ARLAN 2 mb/s.

At this writing, the spectrum allocation proposal seems unlikely to be adopted for regulatory (or some would argue, competitive) reasons, preventing these products from appearing in the European market.

Wave Lan, Radio Link, and ARLAN are all mono-zone products—each workstation contains a LAN card and transceiver and talks directly to other workstations with matching receive/transmit capability, situated within the coverage zone.

An additional option is to use a multi-zone system with a building covered by a number of radio cells, connected by a cabled backbone with wireless "final" links.

Motorola's Altair is a well established, well respected application example in the United States—a TDMA-based system capable of delivering data at 3.5 mb/s and proposed (but not yet accepted) for European implementation at 17.1 to 17.3 GHz.

DECT AS A CORDLESS LOCAL AREA NETWORK

DECT is proposed as a European LAN standard in preference to the existing proprietary solutions described above. There are mixed views as to the suitability of DECT as a fast, dedicated data bearer.

DECT is a cellular system designed to deliver voice traffic and data. It has a call set up and clear down procedure which, its detractors would argue, is optimal for voice traffic and non-optimal for data.

The usual configuration is to have one rf channel allocated per cell, time division divided into 12 traffic channel time slots.

The time division duplexing (TDD) is achieved by subdividing each frame into 24 time slots, 12 of which are used for

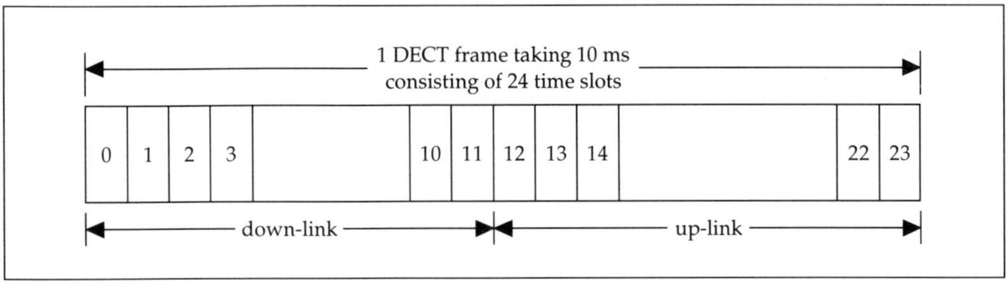

Figure 10.3

the down-link and 12 for the up-link. Each frame lasts 10 ms, providing a throughput rate of 100 duplex frames per second. (See Figure 10.3.)

Each time slot contains 392 data bits, with 320 bits used for information and 64 bits for error checking and control. A frame rate of 100 frames per second implies each user has a duplex net data rate of 32 kb (320 × 100) with 6.4 kb of signaling overhead. (See Figure 10.4.)

The arguments against DECT as a data bearer can be summarized as follows:

1. The 6.4 kb signaling overhead is provided essentially for set up and maintenance (including handover) of voice calls. Each conversation, regardless of whether it is a two-minute voice conversation or a packet of data 600 bits long, needs to be set up using a dedicated control frame and cleared down using another dedicated control frame. Analog cellular, analog trunking, and digital cellular systems share a "comparable constraint" as data bearers; i.e., call set up is optimized for voice as opposed to burst data traffic.
2. DECT's channel coding is optimized for a bit error rate norm of 1 in 1000—adequate for digital voice but disruptive for relatively fast-speed data. A protected data service provides additional error coding and interleaving but at the cost of lower throughput (25.6 kb/s duplex against 32 kb/s).

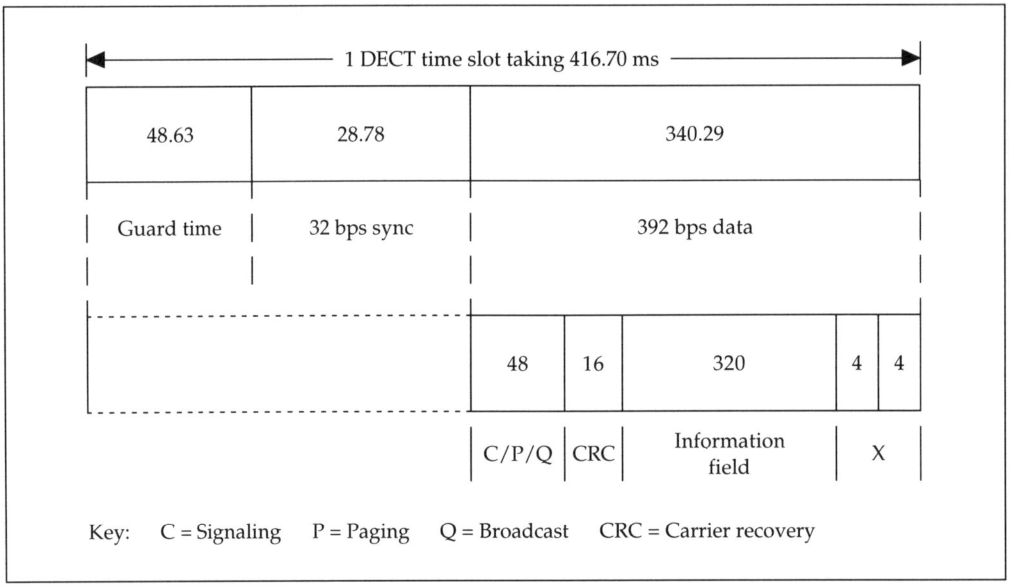

Figure 10.4

3. Time slots can be concatenated to increase data rate; e.g., duplex bearers can be converted to simplex bearers, doubling throughput but with a doubling of set up and clear down overhead. Three channels all made simplex would deliver 153.6 kb/s or an asymmetric link of, say, 128 kb/s (up-link)/25.6 kb/s (down-link).

Call set up and clear down, delivering short data packets reduces throughput per workstation by at least 50%.

The DECT protocol does allow for the allocation of fixed dedicated bearers delivering a 32 kb/s duplex connection (sufficient for E-mail, but not for faster data transfer applications), and for a connectionless protected data mode—a 25.6 kb/s data rate delivering packets of predetermined 512 bit message length. It is arguable whether this is sufficiently flexible or fast to provide the basis for DECT to be the Pan-European LAN standard. The debate continues.

Higher Data Rate Applications

Data rates over 10 mb/s cannot, at this writing, be effectively served by existing radio frequency solutions. Wireless alternatives relay on infrared links, necessarily line of sight.

NOTE The reader is referred to two source articles on cableless LANs and DECT by Douglas Postlethwaite, *Communications International*, September 1991 and *Communications*, October 1991, International Thompson Publishing, United Kingdom.

Wide-Area Networks (Outdoor)

DECT is intended as a localized outdoor; i.e., neighborhood network—for example an outdoor exhibition site but not as a wide-area network—regional or nationwide.

Manufacturers and vendors of portable computers are looking for a delivery mechanism that will allow lap-top, handheld, or notebook PCs to access, interrogate and download to central databases. Such products include AT&T's wireless version of the Safari laptop and notebook PC, NCR's notebook PC, the HP Palm Top 95, and Wang product offerings. Recent announcements include IBM's PC Radio and various Apple proposals.

In the United Kingdom, vendors have the option of using the analog or digital cellular networks or the dedicated data networks—delivering data rates up to 9600 kb/s.

In the United States, vendors have the option of accessing existing analog and digital cellular networks or established packet networks such as Ardis.

IBM's Ardis network, based on Motorola infrastructure and in place since 1985 now supports over 25,000 IBM service engineers with spare capacity sold to Sears, Otis, and New York Life Insurance—using Ardis for live access, wireless, portable PC applications.

Apple Computer is, at this writing, petitioning the FCC for 40 MHz of spectrum between 1850 and 1990 MHz to provide a wide-area wireless access network—an outdoor complement to their indoor wireless version of their Apple LAN, AppleTalk.

NOTE There is a grey area between outdoor and indoor coverage. Most outdoor radio networks are dimensioned to provide a measure of in-building coverage.

The differentiation is that dedicated indoor CLANs are optimized for essentially static but high data rate applications, and digital cellular and dedicated wide-area data networks are optimal for *mobile* applications, where propagation conditions (dynamic multipath) necessarily result in relatively lower rates of data transfer per kHz of utilized bandwidth.

CHAPTER

11

CDMA/SPREAD SPECTRUM

The principle of *spread spectrum* is to transmit information using a far larger bandwidth than the modulation alone requires.

CDMA—code division multiple access—is a technique to overlay many users onto the same segment of spectrum with a minimum of co-user interference.

A PN (pseudo-noise) or user-unique code sequence is added to the information data and modulated onto the carrier. The same PN sequence is used in the receiver to produce a "phase argument" match or "correlation" with the incoming carrier. The correlator only accepts signal energy from the valid binary sequence, despreading the wanted information but spreading other signals with nonvalid PN sequences. Nonvalid signals are therefore "lost" across the channel bandwidth and appear as low-level noise. The required information is correlated, concentrated, and recovered. (See Figure 11.1.)

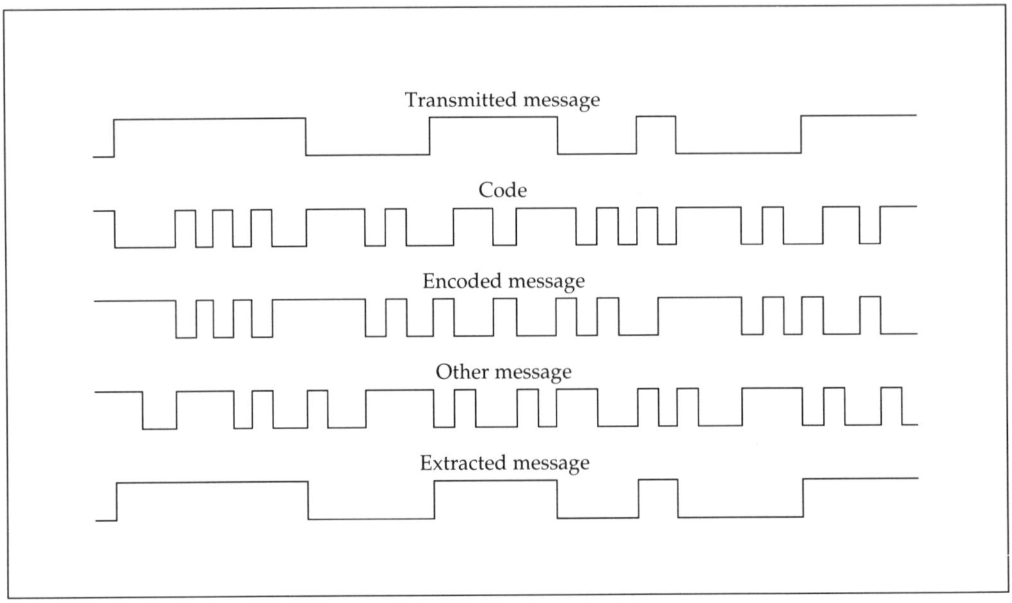

Figure 11.1 *How CDMA works*

The original message is multiplied by the user-unique code. Conceptually you can think of the unique code as performing the function of a "carrier" with a predefined phase argument (waveform property) that can be identified by the receiver. The receiver then subtracts the "carrier" and recovers the wanted information.

CDMA and spread spectrum techniques have been widely proven in military applications for secure anti-jamming communication links and in satellite communication. The test has been to transfer the technology successfully to terrestrial, commercial mobile radio applications. The appeal of CDMA for terrestrial digital cellular is that it transfers much of the complex processing needed at rf in TDMA systems to baseband, provides a soft migration from cell to cell (the mobile receiver gradually achieves a "phase lock" with the next cell), eliminates the problem of multipath signal phase cancellation (by using a bandwidth of at least 1.25 MHz), and allows a CDMA spread spectrum network to coexist with established TDMA networks. For example, one proposal is to take 6 × 200 kHz GSM channels to make a 1.2 MHz underlay supporting a user community of CDMA phones. The CDMA phones would operate well below the noise floor of the GSM network; i.e., would be invisible to GSM users.

As the number of PN sequences used increases, cross-correlation increases between the codes causing co-channel interference and placing an upper limit on the number of users per cell. What this limit is, is a matter of some discussion but the champions of CDMA in the United States, Qualcomm/Nynex, claim a 20-times capacity increase over analog cellular.

Qualcomm trial projects in the United States have used a variable rate vocoder (4.8 kb/s → 16 kb/s) with voice activity and voice intensity detection. The baseband signal is modulated and spread over a bandwidth of 1.25 MHz.

If the spectrum segment is shared and very low noise power is required at any specific frequency, spread spectrum can combine a wide bandwidth with frequency hopping. Omnitracs in the United States using Qualcomm technology provides two-way messaging and reporting to mobile users using geostationary GPS satellites, sharing capacity with Loran C users at 12 to 14 GHz. A low data rate baseband (55 to 165 bps) is spread over 1 MHz by using direct sequence CDMA and then frequency hopped over 48 MHz.

Omnitracs offers the following as a guide to how many users can co-share a segment of spectrum using both techniques.

The number of users N that may simultaneously share the same spectrum using a unique PN sequence is given by:

$$N = \frac{\frac{W}{R}}{\frac{E_b}{N_o}} XF$$

W = Spreading bandwidth

R = Date rate

E_b = Energy per bit

F = Frequency reuse factor

N_o = Noise level—thermal and interference from other users; e.g., for Omnitracs

$F = 48$

E_b/N_o of 5 dB

$W = 1$ MHz

$R = 55$ bps

$N = 32{,}000$

QUALCOMM'S UNITED STATES AND RoW PROPOSALS

Qualcomm, founded in 1985 by Dr. Irwin Jacobs and Dr. Andrew Viterbi, has a widely respected industry track record with CDMA already implemented successfully in Omnitracs (20,000 users) and widely applied in military and satellite applications.

Qualcomm presents a seductive case for the adoption of CDMA as a terrestrial solution, which we will summarize below.

The Variable Rate Vocoder

With a typical voice conversation, only a third of the air time carries speech, two-thirds of the air time is empty space (pauses or listening). Qualcomm has its own version of the CELP (codebook excitation linear prediction) codec—the QCELP, which encodes active speech at 8 kb/s, reducing to 1 kb/s for the coding of the pauses and background noise.

This increases system capacity and delivers a quality benefit by reducing the effect of high noise levels (attenuated by the low bit rate coding).

Multipath Gain

Qualcomm promotes the virtues of its "Rake Receiver." This selects the three strongest multipath signals from among the multiple paths that have a delay spread longer than 0.8 μs (i.e., the delay spread exceeds the PN chip deviation) and combines them to produce one strong signal. At the base station, the four strongest paths received by a dual diversity antenna are correlated and combined.

NOTE CDMA detractors point to the critical need to balance the received signal level from all mobiles accessing one specific cell site—mobiles need to be arriving at the cell site within 1 dB of each other to avoid false correlation.

Soft Capacity Limits

Qualcomm claims that TDMA systems suffer from hard capacity limits. As a cell reaches maximum capacity, there is a rapid increase in call blocking. CDMA can soften this threshold by accepting a higher bit error rate (i.e., lower voice quality), temporarily supporting more users (the traffic peak).

Soft Hand-Off

The CDMA networks use the same channel frequency in all cells. The mobile does not normally change frequency on hand-off; i.e., it stays within the same 1.25 MHz segment on the same center frequency.

As the mobile enters the transition area between cells, it makes a connection to the new cell while maintaining the connection to the old cell, utilizing the best signal paths from the two cells and avoiding signal interruption. This is claimed to deliver a significant improvement in data and fax transfer at the cell boundaries, and avoids the high signaling load imposed on analog or TDMA networks by a mobile "hunting" between cells.

Power Control and Power Budget

A CDMA mobile monitors the power received from the cell site (down-link). The CDMA cell site measures the quality and signal strength of the up-link (mobile to base), power control is more or less continuous (every 1 ms) throughout the call, minimizing interference, reducing fading at slow mobile speeds and minimizing the power budget. (Longer talk-times for a given microportable weight/volume ratio.)

Flexible Data Rates

The appearance of the CDMA-modulated signal does not change if the baseband data rate is changed.

This makes it easy to accommodate non-voice data services (fax or PC traffic) at different data rates or with varying error control requirements. Non-voice data services within CDMA networks will probably be supported by specialized servers at the MTS.

It also makes possible the provision of different grades of speech traffic at different tariffs—for instance premium tariff toll quality, medium tariff communications quality, low tariff synthetic quality.

Migration

CDMA can be introduced gradually into an existing analog network, taking initially 1.25 MHz of spectrum and the further 1.25 MHz segments as required, with users initially having dual mode CDMA/analog AMPS phones, migrating finally to dedicated CDMA cell phones. For a graphic depiction of the comparison of FDMA, TDMA, and CDMA, see Figure 11.2.

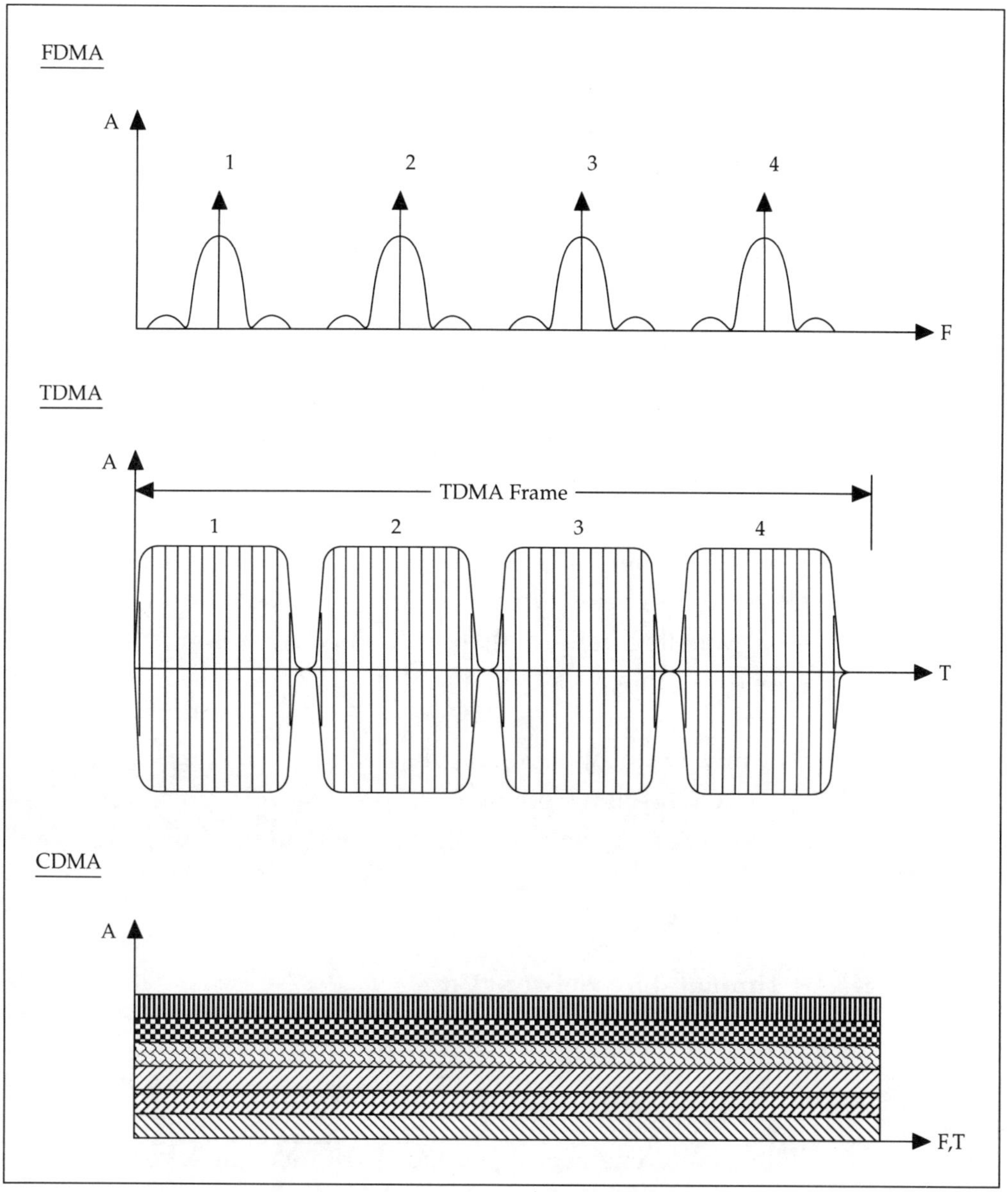

Figure 11.2 *Visual Comparison of Multiple Access Options*

NOTE A CDMA-dedicated network avoids the need for cell-by-cell frequency planning (the same subset of frequencies are used in every cell).

Device Technology

Qualcomm proposes (and has available) a three-ASIC chip set for infrastructure manufacturers and a three-ASIC chip set for mobiles, with one ASIC common to both. Future plans include a one-chip implementation to provide the basis for compact, low-cost, low-power budget microportables.

Validation Trials

The device technology and network is being demonstrated in validation trials in San Diego, to prove that the system capacity and voice quality claims for CDMA can be delivered in real-life conditions.

Assuming, as seems likely, the technology can be shown to have a number of unique advantages over TDMA or narrow-band analog options, there still remains a formidable political lobbying process to convince the CTIA (well down the path of TDMA commercial lock-down) to co-adopt CDMA as a parallel or preferred technology platform.

Qualcomm's politically astute move to introduce CDMA into Korea will help to focus attention on the technology's performance and profit potential.

Through-Life Network Costs

One of the most compelling arguments for CDMA from the carrier (network operator) viewpoint is that it promises to deliver low through-life network support costs (rf engineering operational and engineering support) with a simple, stable rf environment compared to the time domain, frequency domain, and phase domain complexity (at rf) of the TDMA networks; i.e., an enhanced through-life profit profile for the carrier. Note the signficance of this argument for networks with a 15 to 20 year life expectancy.

CHAPTER

12

VOICE QUALITY STANDARDS

In terms of capacity or spectrum efficiency, we can only compare networks as like-for-like if we can demonstrate that the networks deliver comparable voice quality.

In practice, with high bit rate voice coding systems (> 16 kb/s) we can compare performance on the basis of objective signal-to-noise measurements.

As bit rates reduce below 16 kb/s, there is an increasingly poor correlation between signal to noise ratio and the subjective quality of the decoded speech signal.

Voice quality therefore is better compared by using subjective testing—a group of listeners evaluating voice quality and scoring against a mean opinion score in which:

> 4	equates to toll quality
3.5 to 4	equates to communications quality (typical 12.5 kHz two-way radio)
2.5 to 3.5	equates to synthetic quality

Analog AMPS/TACS has a subjective speech quality score of 3.00 with toll quality as a target (not necessarily attained) for digital cellular.

CHAPTER 13

PERSONAL COMMUNICATION NETWORKS—PCN

Personal Communication Networks are a product of the Thatcher decade—a drive to expose telecommunications or mobile communications monopolies or duopolies to broader competition.

Three licensees, Microtel, Unitel, and Mercury PCN have committed to build three national mobile communications networks at 1.7109 → 1.880 GHz, each using 50 MHz of spectrum with 20 MHz duplex spacing. Unitel is working with Mercury PCN on a "Parallel Network Architecture" network sharing proposal.

PCNs use the same network architecture as GSM with base transceiver stations (BTS) radio sites, base station controllers (BSC) (network sites), and a mobile switch centre (MSC).

BSCs can support one to 32 base transceiver stations—equivalent to the maximum number of control channels, as in GSM.

The networks have the option to use 38 GHz microwave/millimetric line-of-sight links for 2 mb traffic up to a range of 18 km, or PCM 30 leased lines.

PCN in the United Kingdom is based on GSM technology but optimized for a hand-portable (microportable) user community using low-power mobiles supported by small cells. The adoption of GSM technology reduces development cost. Implementing the network at 1.8 GHz means more users per MHz of spectrum (higher propagation loss, therefore more frequency reuse) but means higher infrastructure cost—more base stations needed for equivalent geographic coverage.

Because cell sizes are small, handset power can be reduced, reducing battery drain and/or improving talk-time and/or reducing size and/or weight. The PCN power budget gives an idea of the up-link and down-link requirements.

Up-link (Mobile to Base Station)

Transmit power	24 dBm Peak
Transmit antenna gain	–3 dB
Path loss	133 dB
Fade margin	7 dB
Receiver antenna gain	15 dB
Receiver sensitivity	–104 dBm

Down-link (Base Station to Mobile)

Transmit power	38 dBm Peak
Transmit antenna gain	15 dB
Path loss	133 dB
Fade margin	7 dB
Receiver antenna gain	–3 dB
Receiver sensitivity	–90 dBm

Most equipment being developed for GSM can be used with very little modification except for the rf front-end. PCN does have the advantage that the lower power requirement obviates the need for high-power amplifiers.

The ETSI specification for PCN has been developed into the DCS 1800 standard based on a number of delta recommendations that include a common air interface between the networks and relaxed performance requirements—125 km/h maximum operational speeds and smaller maximum cell radius. Mobile power is from 0.25 watt to 1 watt, with base stations from 2.5 watt to 30 watt.

PCN PROPAGATION

Most of the propagation research for PCN at 1.8 GHz has focused on the modeling of typical "urban" conditions and the effect of multipath on a transmit burst; i.e., the impulse response of the channel. Impulse responses typically are measured across a delay interval and/or a delay window. A delay interval is the length of the impulse response above a specified amplitude threshold, the delay window is the length of the power profile containing a certain percentage of the total energy found in the impulse response. (See Figures 13.1 and 13.2.)

Propagation modeling up to 1 GHz is often based on the HATA model, not directly applicable to 1.8 GHz because of the increased importance of diffraction edge losses, particularly significant with antennas mounted at or below roof height. Work by Walfish, Bertoni, and Ikegami modifies the HATA model with a roof-top loss factor and a street-orientation factor.

Although diffraction edge losses and free space losses increase, the other major source of signal loss—signal phase cancellation from multipath, can be controlled to an extent by

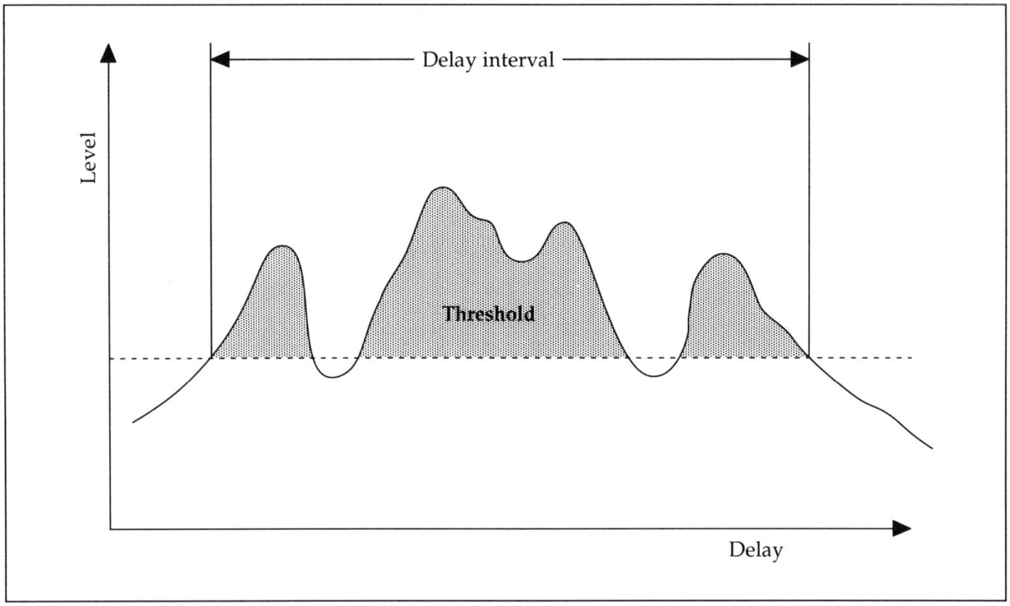

Figure 13.1

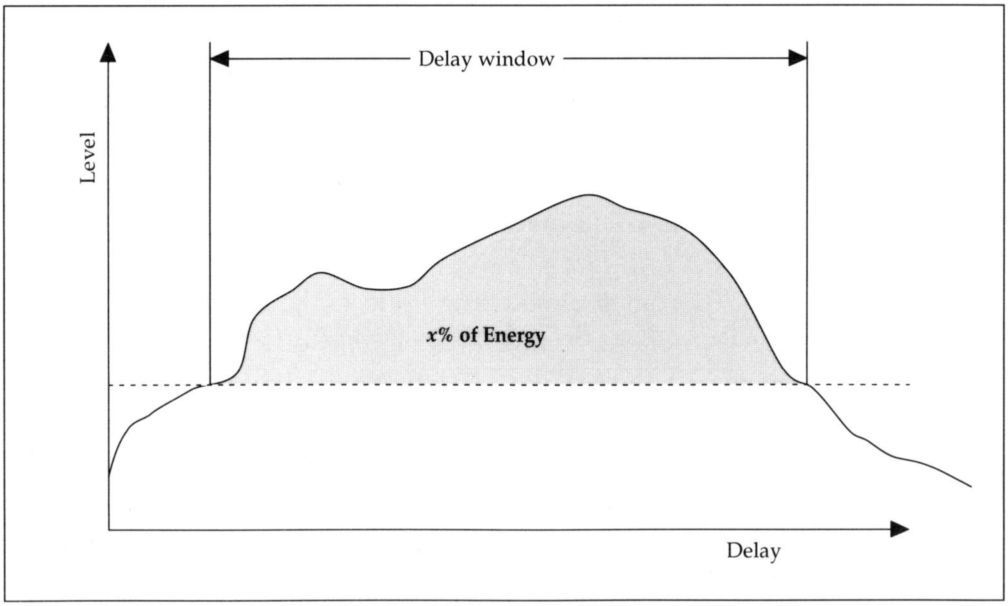

Figure 13.2

implementing diversity reception (shorter wavelength means reduced, i.e., more convenient, diversity spacing).

One objective of the PCN operators has been to implement a system that can replace the local loop into buildings—from pavement to PABX or domestic phone. The building penetration characteristics at 1.8 GHz have been the focus of considerable research work published by CCIR Study Group 5 (Report 547). In practice, buildings vary enormously in terms of their rf propagation characteristics. An initial path loss of 10 dB to 30 dB into the building, an accumulated loss through each partition wall of 0.4 to 28 dB depending on wall construction (metallic backed plasterboard, for example). Buildings (commercial and domestic) are notoriously nonstandard and defy simple modeling. The rf problem has a direct parallel with heat-loss and heat-gain calculations. To calculate a heat-loss or heat-gain profile, each component part of the path has to be analyzed in terms of its heat transfer characteristic. In the United Kingdom, the Chartered Institute of Building Services (CIBS) publishes an enormously long book of common building materials—a parallel publication will be needed to support accurate rf calculations. Penetration characteristics change with building height—generally improving with height gain (clearer line of path to the base station antenna). Unpredictable effects can upset theoretical calculations—for example triple glazing with the glazing sandwich at a wavelength resonance, can act as a waveguide (the rf becomes trapped between the panes). Lift shafts can act as waveguides. Figure 13.3 illustrates an assumed low loss path and two actual paths via two windows and a reflection point and three walls and an elevator shaft.

The reader is referred in particular to the work of Professor David Parson's Communications Research Group at the University of Liverpool in the United Kingdom on 1.8 GHz propagation research.

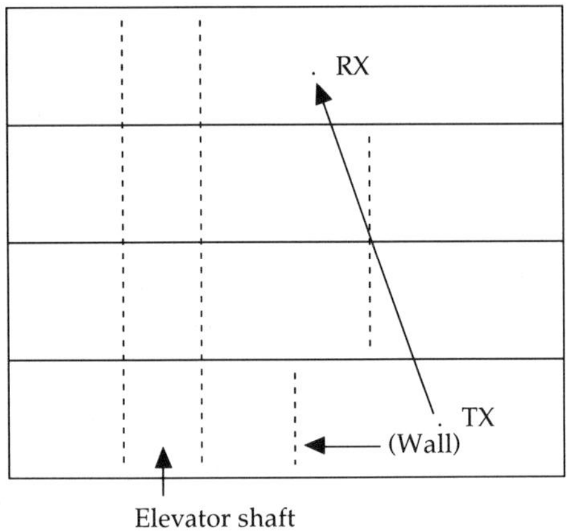

(a) Assumed direct path through three floors and one wall

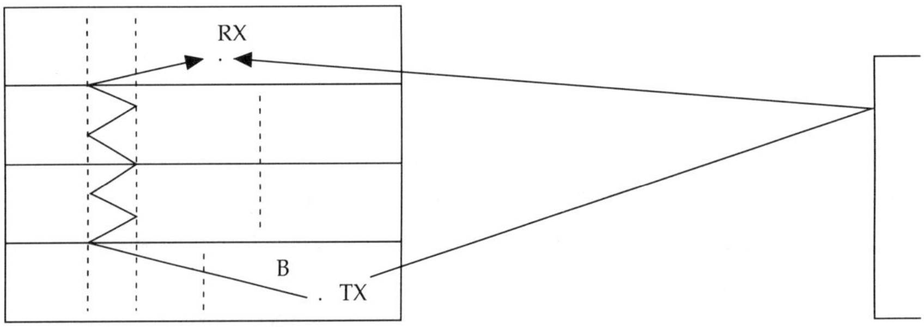

(b) Alternative low-loss paths
(a) Via two windows and one reflection point.
(b) Via three walls and elevator shaft.

Figure 13.3 *Assumed direct path through three floors and one wall*

PCN ECONOMICS

What is very obvious is that to offer directly equivalent geographic coverage at 1.8 GHz at the specified power output levels will require three to four times more base stations than the existing 900 MHz analog and 900 MHz digital networks—i.e., between 2,500 and 3,000 base stations per operator. Figures of

Table 13.1

<table>
<tr><th></th><th></th><th>Traffic</th><th>Revenue</th><th colspan="2">Profit</th></tr>
<tr><td>LOCAL</td><td rowspan="3">Competition including PCN Access</td><td>80%</td><td>35%</td><td>Min or loss</td><td rowspan="3">Competition including PCN</td></tr>
<tr><td></td><td></td><td></td><td></td></tr>
<tr><td>TRUNK</td><td>18%</td><td>35%</td><td>Median</td></tr>
<tr><td colspan="2"></td><td></td><td></td><td></td><td></td></tr>
<tr><td colspan="2">INTERNATIONAL</td><td>2%</td><td>27%</td><td>High</td><td rowspan="2">Duopoly Mercury British Telecom</td></tr>
<tr><td></td><td></td><td></td><td></td><td></td></tr>
</table>

£1 billion have been mentioned to build, implement, and load each network.

This takes substantial financial commitment and explains why the PCN players have to be backed by substantial companies; Microtel for instance backed by Hutchison (£1.47 billion turnover in fiscal 1990 mainly from Hong Kong) and Mercury PCN backed by Cable and Wireless (£2.6 billion turnover with 60% of profits from Hong Kong). City analysts refer to the "Tiananmen Square effect"—the need to move cash into capital-intensive "safe haven" United Kingdom projects with long-term revenue potential.

The fiscal risk of PCN is balanced against the long-term goal of capturing revenue from the POTS (plain old telephone service). Of British Telecom's £2.5 billion revenue, over 70% is from POTS with a traffic/revenue/profit split shown in Table 13.1.

NOTE BT average revenue from residential subscribers is £100. BT average revenue from business subscribers is £675. Cable and Wireless (Mercury) currently have 4% of BT's revenue. They see Mercury PCN as strategically necessary to increase market share exponentially, with the longer term aim to re-export the local loop bypass expertise back to the Pacific Rim. The trick will be to successfully extract profit (as opposed to revenue) from the domestic customer base and build market share in the business sector and international traffic market.

RACE PROGRAM AND THE UMTS

Reference should be made to the RACE program—Research into Advanced Communications in Europe—and the UMTS concept—Universal Mobile Telecommunication System. "UMTS combines the wider area roaming and coverage capabilities of cellular systems with the high density and quality capabilities of cordless systems": a quotation from the UMTS mission statement. UMTS is intended to be IDSN compatible and based on a microcell structure combining the virtues of all existing systems and the vices of none. Data rate will be up to 2 mb/s, with a bandwidth allocation of approximately 200 MHz around 2 GHz.

Watch this space. (For some time.)

NATIONAL TELEPHONE SERVICES BASED ON FIXED WIRELESS COMMUNICATIONS

This is reality in some countries—for example Malaysia and parts of Scandinavia where analog NMT450 systems provide a more economical way to provide telephone service to remote populations than by provision of land lines.

NOTE A fixed cellular phone can have a high-gain antenna permanently fixed on a roof top.

For urban populations, fixed wireless has to compete with existing copper-wired systems (many of the networks amortized over 30 to 50 years) and fiber to the pavement. A United Kingdom champion of fixed wireless telephones has been the company Ionica L3.

U.S.A. PCS (PERSONAL COMMUNICATION SYSTEMS)

PCS or PCN always run the risk of becoming a political football, particularly when national telecommunications providers become threatened. Many interesting opportunities exist to exploit underutilized or fully amortized resources. AT&T in the United States, for example, has a microwave network of 3000 masts, which it would like to use to support a national user community of mobile 6-GHz microportable pocket phones.

CHAPTER

14

SATELLITES

HISTORY

The first United States rocket was launched in 1926—500 years after the Chinese invented gunpowder. The Germans helped to perfect the rocket maker's art, building V1 and V2 missiles in the Second World War. Much of this expertise was not lost post-war but transferred to the United States. By 1945, Arthur C. Clarke was predicting a network of "extra terrestrial relays" spanning the world. In the 1950s, the space race (Sputnik and Explorer) nudged launch technology forward a notch or two. By 1962, Telstar was providing telecommunications service. By 1965, President Johnson was acclaiming Intelsat's Early Bird as "A milestone in the history of communications between people and nations." Between 1966 and 1969, the Apollo deep space missions had a profound influence on the technology used in the launching and tracking of satellites and satellite communication capability.

Satellites suffer from politics (they travel across national boundaries) and Murphy's Law (they go wrong). Two famous (infamous) examples include INSAT, the Indian telecommunications satellite whose earth sensor locked on to the moon during an eclipse in 1982, and the European Olympus satellite whose (Italian) controller accidentally turned the satellite's solar panels away from the sun in summer 1991—a satellite deep freeze in the sky, now slowly thawing.

GEOSTATIONARY SATELLITES

For mobile applications, geostationary satellites can be very suitable for one-way paging or for positioning (e.g., Inmarsat M services). For the higher latitudes, south and north (e.g., Northern Europe), the low elevation path to a geostationary satellite can result in blocking from terrain and buildings. For two-way applications, the main constraint inhibiting the use of geostationary satellites is the long link path, including a long link path through the atmosphere resulting in a high up-link power budget and long path link delay.

HIGH ELLIPTICAL-ORBIT SATELLITES

An alternative for Northern Europe has been to consider the Molniya (Russian for *lightning*) or Tundra highly elliptical orbits. The Molniya orbit would use three satellites 120° apart, each satellite active eight hours every day. A problem with the Molniya orbit is that it passes through the Van Allen belt near its perigee, implying significant risk of damage.

The Tundra orbit would have two satellites each giving 12 hours of coverage. One problem is the apogee distance—between 46,000 and 53,000 km—producing a one-way delay of

350 ms, and Doppler effects implying path correction processing overhead.

Either orbit does have the advantage of high elevation over Europe. Application studies have considered a 40 beam spot satellite giving reconfigurable coverage across 200 to 300 kilometre footprints. The delay budget however, would limit the usability or acceptability of the network as a speech bearer.

LOW-ORBIT SATELLITES

Motorola is proposing a low-orbit satellite system to support mobile radio users in remote regions worldwide.

NOTE Satellites are good at providing wide-area coverage, for instance over mountains or rural areas where it is expensive to service the user community with a terrestrial network.

Motorola's proposal is to launch and maintain 77 satellites in seven equally spaced orbits over the North and South poles at an orbiting height of 600 km. Inmarsat, Canadian Telesat, and American Mobile Sat have expressed interest in backing or being involved with the project, named *Iridium* (the 77th element).

Each satellite would have a 3000-mile footprint addressed by 37 narrow beams. The satellite output power would be 1000 times lower than that needed for geostationary satellites, reducing size and weight and (therefore) launch costs. Launch costs will be critical since, even assuming a five-year (optimistic) in-orbit life expectancy, a replacement launch would be needed every 24 days:

$$\frac{5 \times 365}{77} = 24$$

Motorola is proposing a "warm" production line—i.e., continuous production—to reduce costs per satellite. The antennas are fixed phased arrays, avoiding the need for deployment; i.e., the solar packs are the only part of the satellite that will need to be deployed in orbit, meeting size and weight constraints, which will make a Pegasus launch or multiple Ariane or Delta launches possible. Motorola claims an on-orbit cost per satellite of less than $30 million.

Iridium has the capability of being a bearer for international long distance traffic—the most profitable sector of worldwide telecomms. Using fast intersatellite switching at 60 GHz to deliver traffic country to country avoids the long (260 ms) delays of geostationary links, providing a plausible competitor to fast fiber-optic international links.

Mobiles could display time, latitude, longitude, and altitude. Power budget on the up-link will be approximately 7 watts (just below the biological safety margin). Voice coding will be the 8 kb Motorola codec, and the proposal is to include a data bearer at 2400 bps.

Motorola is bidding for a 10 MHz frequency allocation between 2 GHz and 3 GHz at the 1992 WARC (world administrative radio congress). The complexity of the Iridium proposal can be appreciated by considering the separate segments.

The space segment—the 77 satellites

The earth segment—20 GHz and 30 GHz up-link earth stations for delivery of traffic from national PSTNs or national networks for onward delivery on Iridium cross-links and down-links.

The mobile segment—mobile up-links and down-links. To support low-power, hand-held terminals with low-gain antennas, the satellite spot beams must be able to adapt power levels to the link distance from satellite to mobile.

Using a 7-cell frequency reuse pattern, cells are scanned by the satellite antenna arrays with a specific time pattern and sequence. Each of the satellite scanning beam antennas is programmed to aim the correct cell at the earth at a set time.

Subscriber units accessing the cells or being addressed by the cells will weigh less than 0.6 kg with 24-hour operation on standby.

Questions still remain as to how many people will be prepared to pay for a premium-priced service in rural areas or for use in extreme and exceptional operating conditions. For the more specialist applications, there are competitive, lower cost, albeit lower quality, speech and data bearers such as HF (short wave) or meteor scatter systems. Iridium will also meet substantial political opposition from national telecommunications providers jealous of their transnational traffic revenue base.

Space Systems Loral (previously Ford Aerospace) has filed a simpler submission to the FCC for a 45 satellite system with no intersatellite switching—politically more acceptable internationally.

CHAPTER

15

MILLIMETRIC WAVELENGTHS

RESONANT AND NONRESONANT ABSORPTION

As a general principle we can state that as frequency increases, bandwidth increases—a prime reason for moving to new "green field" parts of the spectrum. Millimetric wavelengths span the frequency spectrum from 30 GHz to 300 GHz below the black body curve, infrared, and light. Above 30 GHz, the wavelength becomes closely equivalent to the dimensions of a raindrop, resulting in a propagation characteristic of nonresonant absorption—scatter by raindrops, hail, and snow.

This is a vice not a virtue, implying a degree of unpredictability on the path link. In practice it rains for only a small percentage of the time—only 4% or 15 continuous days per year in the United Kingdom (people never believe this) and snow and hail are *very* rare (statistically speaking).

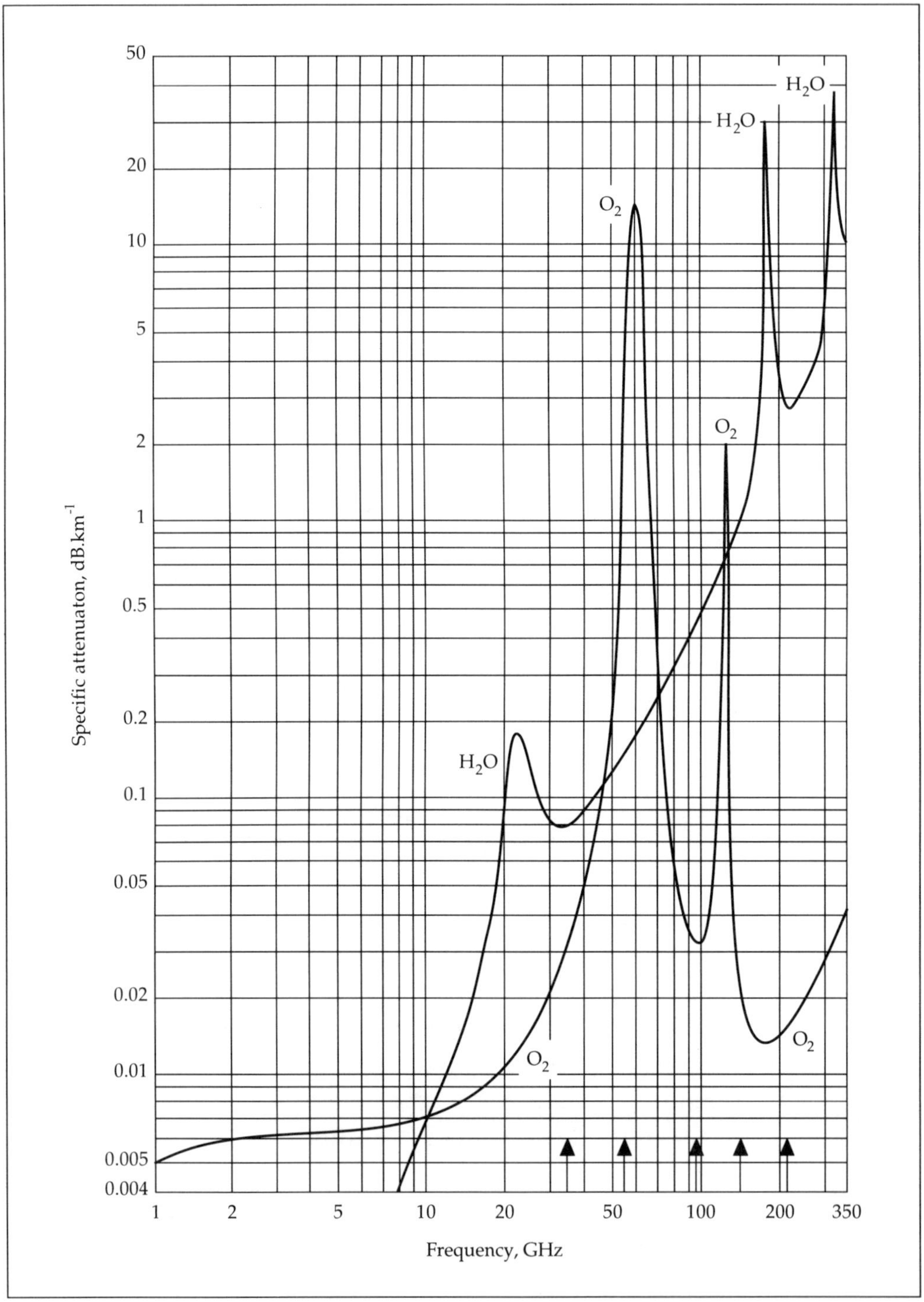

Specific attenuaton, dB.km^{-1}
50
20
10
5
2
1
0.5
0.2
0.1
0.05
0.02
0.01
0.005
0.004
1
2
5
10
20
50
100
200
350
Frequency, GHz
H_2O
H_2O
O_2
O_2
H_2O
O_2
O_2

Figure 15.1

Millimetric wavelengths are also subject to "resonant absorption" at specific frequencies, caused by molecules of oxygen or water vapor.

There is an attenuation peak for water vapor at 22 and 183 GHz and an attenuation peak for oxygen at 60 and 119 GHz. (See Figure 15.1.)

THE 60-GHZ OXYGEN PEAK

The 60-GHz oxygen attenuation peak is of particular interest for mobile radio applications, giving an attenuation of 15 dB per 100 meters. In practice, this means that the rf coverage from a transmitter is very similar to the lux level lighting coverage for a street lamp. Lighting designers and rf propagation designers both use polar diagrams (such as those shown in Figure 15.2, next page) to show levels of intensity and coverage limits.

THE LAMP-POST LAN (LOCAL AREA NETWORK)

So we have a part of the spectrum that has plenty of available bandwidth; i.e., able to support broadband high-speed data, voice or video applications, with a propagation characteristic that allows us to reuse frequencies from street to street. We have the concept of a *lamp-post LAN*. A lamp-post LAN is a street lamp with a small rf transmitter incorporated either into the lamp housing or into the light itself (high-pressure or low-pressure sodium or mercury). It irradiates the street with rf, providing voice communication to the pedestrian community. It irradiates the houses or offices along the street with rf, providing a broadband two-way link for speech, video, and data. (The local loop by-pass). Street lamps have several benefits: (1) They are already there and do not need planning permission to be there; (2) They are supplied with electricity; (3) They have established installation and maintenance arrangements; (4) They have been placed to provide defined coverage patterns (of light) and are

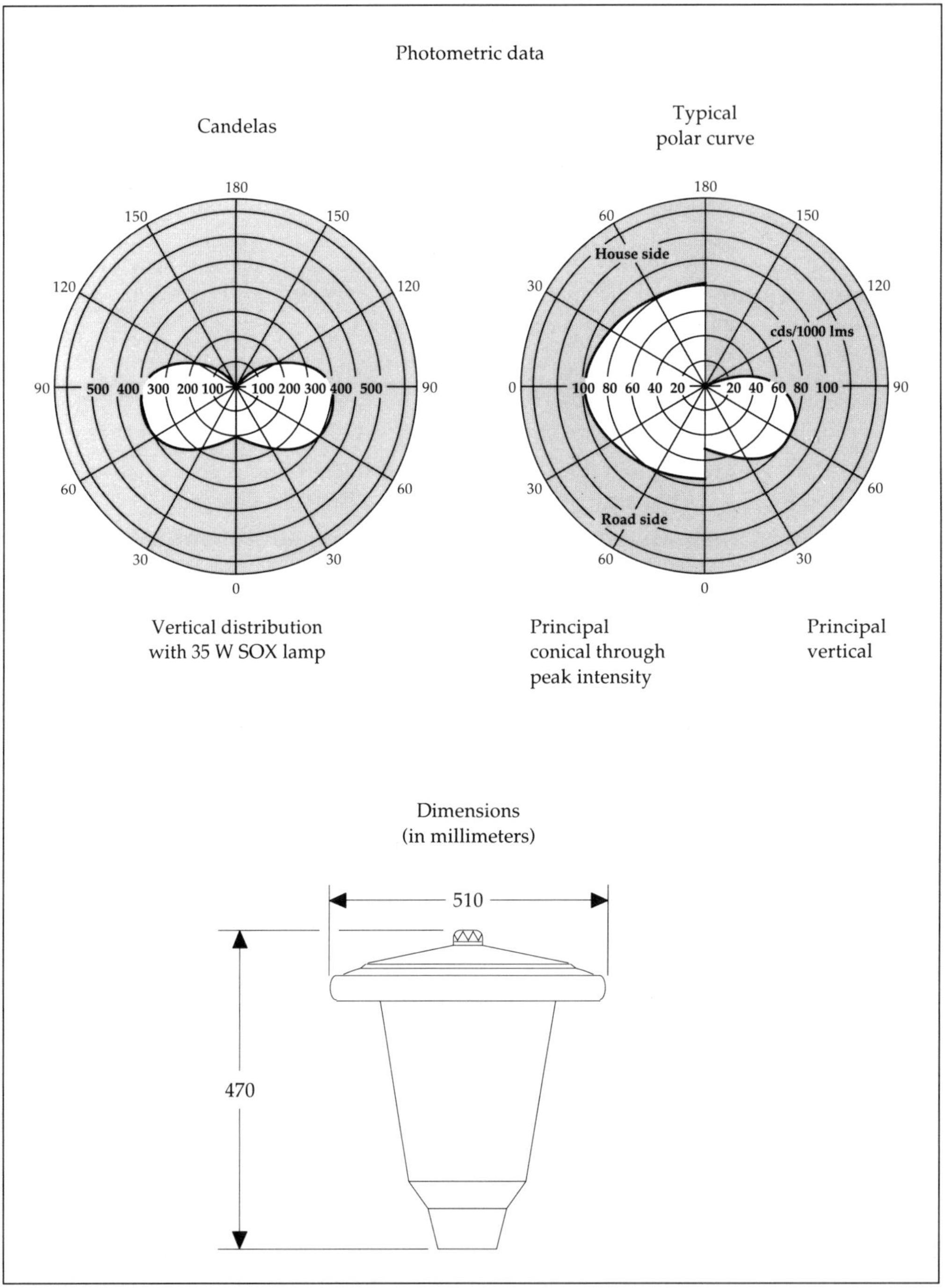

Figure 15.2 *Public precinct lantern*

generally equidistant from one another, simplifying rf propagation planning.

Most main roads and motorways in Europe now have catenary lighting along the central reservation—every catenary lighting module could contain a base station.

Using CDMA spread spectrum techniques, in-car mobiles traveling rapidly from cell to cell (lamp-post to lamp-post) would progressively correlate from cell to cell to provide continuous good-quality coverage.

The lamp-post LAN gets our vote for an elegantly appropriate technical solution to providing broadband two-way mobile communications in the early years of the twenty-first century.

Note also that the use of CDMA technology, translating complex processing from rf to baseband, together with the low-power operating environment, will help realize 60 GHz operation without the use of exotic or expensive device technology.

CHAPTER

16

CONCLUSION

We suggest there will be several competing mechanisms for delivering data, including digital speech and video, in the twenty-first century including fixed copper wire, fiber optic, fixed wireless, and mobile, with mobile served by terrestrial digital macrocellular, microcellular (outdoor), picocellular (indoor), and extra-terrestrial low-orbit systems.

The implementation of new network technology is a "slow-burn" process. It took 20 years for analog cellular systems to translate from the Bell Core Laboratories in the 1960s to the market in the 1980s. Analog cellular networks are still being installed worldwide today with a 20-year life expectancy; i.e., analog cellular systems will have had a lifespan of some 50 years.

It has taken 10 years of specification and committee decision making to take GSM from concept to commissioning, and GSM could have a 40-year lifespan.

In contrast, a new generation of cell phones is introduced every year. The performance of the network is continuously upgraded by improvements in the performance of the mobiles used in the network.

Software engineering is also becoming progressively more important. Networks can gain intelligence through life—analog AMPS and TACS are prime examples.

Analog processing still has considerable development potential. Narrow-band AMPS and narrow-band TACS deliver first-rate quality speech and service levels and spectral efficiency by using high-selectivity filters, superlinear, power-efficient PA chains, and efficient noise-suppression circuits.

The raw processing power of digital devices will however, in the longer term deliver benefits to the manufacturer, network operator, service provider, and user that will be hard to ignore.

The transition or migration process may take longer than popularily assumed. Remember that in 1918, Major Edwin Armstrong developed the concept of the superhet and FM modulation—a technology delivering substantial quality and performance benefits to the user community and substantial competitive advantage to the carriers (broadcasters), but it took 21 years for Armstrong to "leverage" the technology into the marketplace—at his own expense. It took over 50 years for the technology to achieve market dominance, and the two technologies of AM and FM still co-exist today.

In the longer term, analog will move aside to take a gentle retirement. We will regret its passing and miss its elegant simplicity but, next time you take your dog for a walk, take a look at your local lamp-post. Is that really the future for mobile radio?

APPENDIX A

CHARACTER CODES

Table A.1 *ASCII: American National Standard Code for Information Interchange*

b_7						0	0	0	0	1	1	1	1
b_6						0	0	1	1	0	0	1	1
b_5						0	1	0	1	0	1	0	1
Bits	b_4	b_3	b_2	b_1	**Row**	**Column**							
						0	**1**	**2**	**3**	**4**	**5**	**6**	**7**
	0	0	0	0	0	NUL	DLE	SP	0	@	P		p
	0	0	0	1	1	SOH	DCI	!	1	A	Q	a	q
	0	0	1	0	2	STX	DC2	"	2	B	R	b	r
	0	0	1	1	3	ETX	DC3	=	3	C	S	c	s
	0	1	0	0	4	EOT	DC4	5	4	D	T	d	t
	0	1	0	1	5	ENQ	NAK	%	5	E	U	e	u

b_7						0	0	0	0	1	1	1	1
b_6						0	0	1	1	0	0	1	1
b_5						0	1	0	1	0	1	0	1
Bits	b_4	b_3	b_2	b_1	**Row**	**Column**							
						0	**1**	**2**	**3**	**4**	**5**	**6**	**7**
	0	1	1	0	6	ACK	SYN	&	6	F	V	f	v
	0	1	1	1	7	BEL	ETB	'	7	G	W	g	w
	1	0	0	0	8	BS	CAN	(	8	H	X	h	x
	1	0	0	1	9	HT	EM	)	9	I	Y	i	y
	1	0	1	0	10	LF	SUB	*		J	Z	j	z
	1	0	1	1	11	VT	ESC	+	;	K	[	k	{
	1	1	0	0	12	FF	FS	,	<	L	\	l	|
	1	1	0	1	13	CR	GS	-	=	M	]	m	}
	1	1	1	0	14	SO	RS	.	>	N	^	n	~
	1	1	1	1	15	SI	US	/	?	O	_	o	DEL

CHARACTER REPRESENTATION AND CODE IDENTIFICATION

The standard 7-bit character representation, with b_7 the high-order bit and b_1 the low-order bit, is shown below.

Example

The bit representation for the character "K," positioned in column 4, row 11, is

b_7	b_6	b_5	b_4	b_3	b_2	b_1
1	0	0	1	0	1	1

The code table for the character "K" may also be represented by the notation "column 4, row 11," or alternatively as "4/11." The decimal equivalent of the binary number formed by bits b_7, b_6 and b_5, collectively, forms the column number, and the decimal equivalent of the binary number formed by bits b_4, b_3, b_2 and b_1, collectively, forms the row number.

APPENDIX B

3 dB AND 6 dB COMPARISON

6 dB PER OCTAVE LOSS IN DISTANCE

Every doubling of distance results in a

halving of voltage (field strength)
halving of current
a quarter of the power (3 dB is therefore the half-power point).

The term octave is normally used to refer to a doubling of the frequency.

A decade is a 10-times increase in the frequency.

APPENDIX C

THE ERLANG

One Erlang is the traffic intensity in a channel that is continuously occupied.

A channel occupied for 50% of the time carries a load of 0.5 Erlang.

APPENDIX

D

USEFUL COMPARISONS

dBm—dB relative to a milliwatt into 50 ohms.

Table D.1

	Power	**RMS Volts; e.g., Received signal strength**
+ 13 dBm	20 milliwatts	1 volt
0 dBm	1 milliwatt	0.22 volt
–13 dBm	50 microvolts	50 millivolts
–47 dBm	0.02 μ microwatt	1 millivolts
–73 dBm	50 picowatts	50 μ microvolts
–113 dBm	0.005 picowatt	0.5 μ microvolt
–126 dBm	0.0002 picowatt	0.1 μ microvolt

- * mW: milliwatts, thousandth of a watt
- * mvolts: millivolts, thousandth of a volt
- * μwatts: microwatts, millionth of a watt
- * μvolts: microvolts, millionth of a volt

APPENDIX E

DIGITAL MODULATION—

Power Efficiency/ Spectral Efficiency Definitions for Mobile Radio

Digital modulation is used in fixed-link applications to provide for signal regeneration at each repeater; i.e. regenerative repeaters.

NOTE In analog systems, the transmission noise and interference as well as the signal are amplified and retransmitted at each repeater stage.

In digital systems the signal is reconstituted, minus any noise and interference before retransmission. In mobile applications, digital processing is used to correct for multipath effects and allow networks to operate at (relatively) low carrier-to-interference ratios.

In fixed links, higher level modulation methods are used to increase spectral efficiency. In mobile links a delicate compromise is required. In general, it is a requirement that adjacent channel power levels should be at least –60 dB down to avoid adjacent channel interference. (Particularly when digital channels are operating adjacent to analog channels). Power budgets on mobiles also usually influence designers to use power-efficient—i.e., implicitly nonlinear—power amplifiers.

In fixed-link applications, the definition of "power efficient" modulation is to achieve a bit error rate of $< 10^{-8}$ at an E_b/N_o level of < 14 dB (note: E_b/N_o is the ratio of bit energy to noise power spectral density); for linear systems in the presence of AWGN, QPSK is a "power efficient" modulation method. We shall look at ways of realizing QPSK and the compromises required to achieve an acceptable power/spectral efficiency trade-off for mobile radio applications.

QPSK SYSTEMS

In a QPSK modulator, the incoming unipolar binary bit stream is converted into two bit streams (in-phase and quadrature) by a serial-to-parallel converter. It is then converted by two unipolar-to-polar convertors to form two bipolar binary signals. The signals are then passed through two spectral-shaping low-pass filters before being modulated by an in-phase and a quadrature phase oscillator. These two signals are then combined to give a quadrature phase signal. See Figure E.1.

The QPSK signal is then filtered at the combined output to limit the bandwidth and hence prevent adjacent channel interference. It also removes out-of-band spurious signals generated by the modulation process.

In QPSK systems, the modulated signal has four phases, each representing one symbol, each symbol containing two bits of information, usually mapped onto the symbols using a Gray

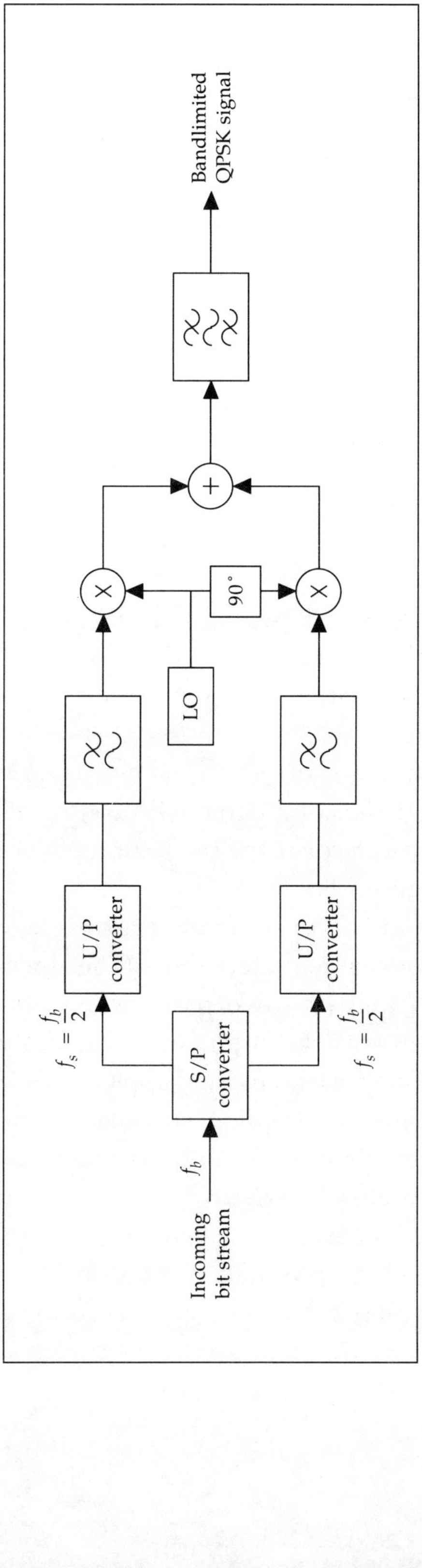
Incoming bit stream
f_b
S/P converter
$f_s = \frac{f_b}{2}$
U/P converter
U/P converter
$f_s = \frac{f_b}{2}$
LO
90°
Bandlimited QPSK signal

Figure E.1

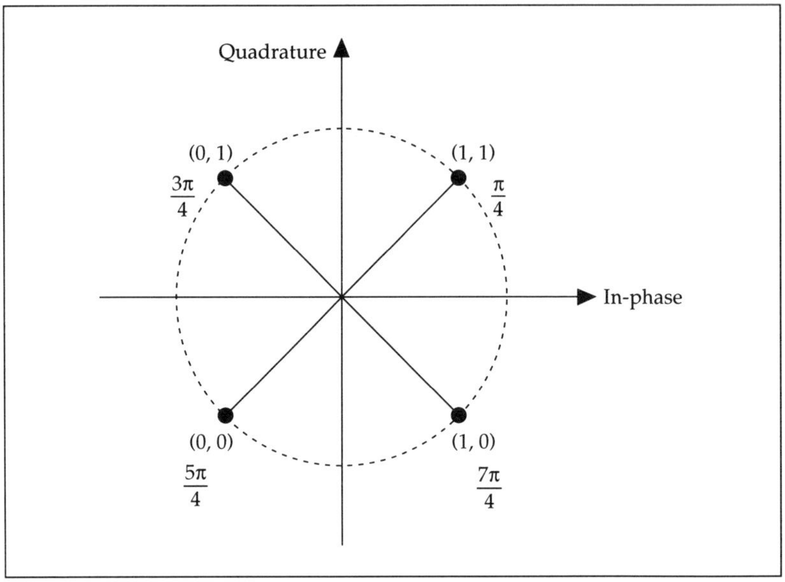

Figure E.2

code (Gray code only permits one bit value change per symbol; i.e., 00 to 11 would be an invalid transition).

The constellation of a Gray coded QPSK signal is as shown in Figure E.2.

Demodulation of QPSK is shown using a coherent demodulator. Due to the nondifferential encoding of the modulation it is necessary to recover the phase of the carrier for demodulation. See Figure E.3.

An input band-pass filter removes the out-of-band noise and adjacent channel interference. The signal at the output of the filter is split into two parts, each part being coherently demodulated with the in-phase and quadrature recovered carriers, then low-pass filtered and applied to one bit A–D converters to regenerate the in-phase and quadrature baseband signals prior to recombination through a parallel-to-serial convertor to deliver the orginal bit stream.

The carrier recovery circuit is used to recover the frequency and phase information of the carrier prior to demodulation. The symbol timing circuit must recover the timing information needed to A–D convert the received signal.

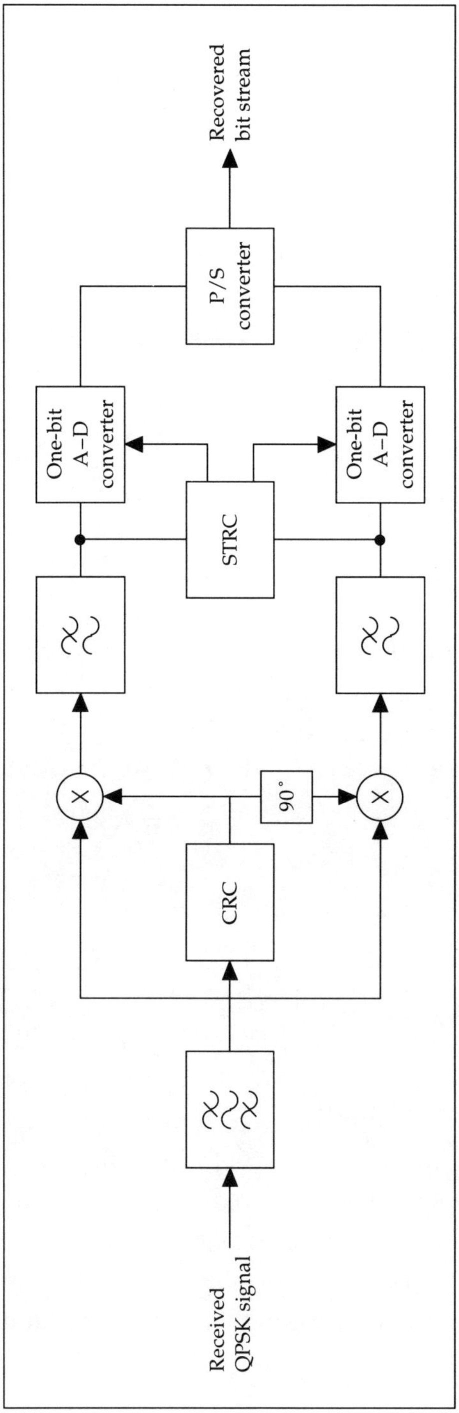
Received
QPSK signal
CRC
90°
STRC
One-bit
A–D
converter
One-bit
A–D
converter
P/S
converter
Recovered
bit stream

Figure E.3

DQPSK

Most carrier recovery circuits introduce phase ambiguity into the recovered carrier. This means that the demodulated "0" and "1" states may be reversed. The baseband data can be processed to remove this resulting ambiguity by using differential encoding and decoding—detection of a (relative rather than absolute) state change indicated by a symbol transition, hence the description of DQPSK, differential quarternary phase shift keying.

QPSK SPECTRAL EFFICIENCY

The spectral efficiency of a modulation scheme is defined as the ratio of bit rate to bandwidth in bits/s/Hz.

The theoretical spectral efficiency of QPSK is 2 bits/s/Hz. A practical filter with a roll off of $\alpha = 0.2$ delivers a spectral efficiency figure of 1.66 bits/s/Hz.

$$\frac{2}{1.2} = 1.66$$

Carrier Recovery Techniques

One simple method for carrier recovery of QPSK is the *Fourth Power Loop*. The received signal is filtered then passed to a non-linear fourth power law device, which generates a spectral component at four times the transmitted carrier frequency from the continuous power spectral density of the received signal. The component is acquired by a conventional phase-locked loop, and the in-phase and quadrature references are obtained using dividers and phase shifters. In this application the phase locked loop is performing the function of a narrow band filter. See Figure E.4.

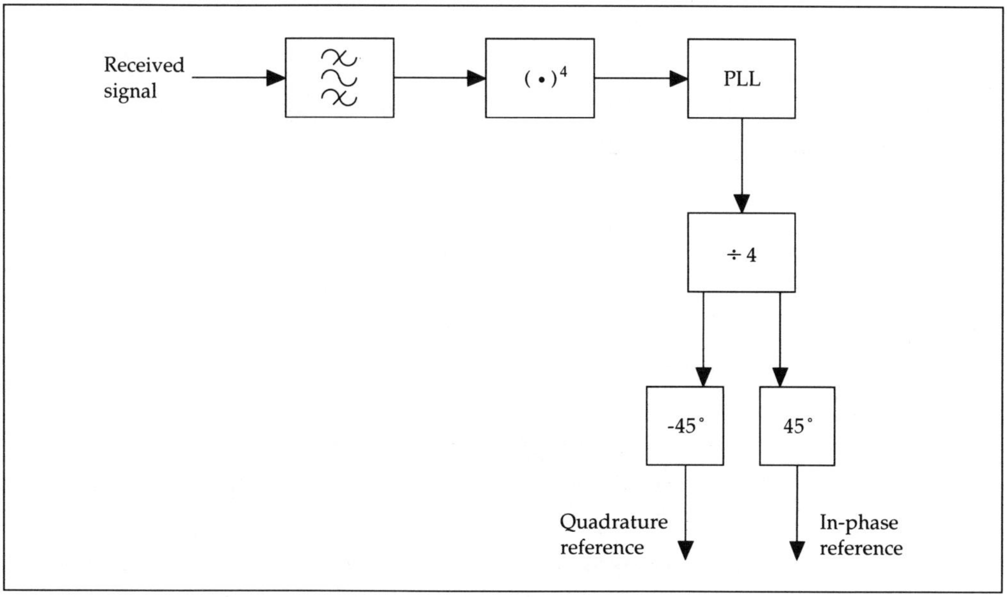

Figure E.4

Timing Recovery

This requires recovery of the bit rate information—i.e., the generation of a discrete spectral component corresponding to the bit rate.

There are two common methods, both using nonlinear operations:

1. Filter and Square Technique

The demodulated baseband signal is low-pass filtered and a square law device is used to generate a spectral line at the symbol rate frequency. See Figure E.5.

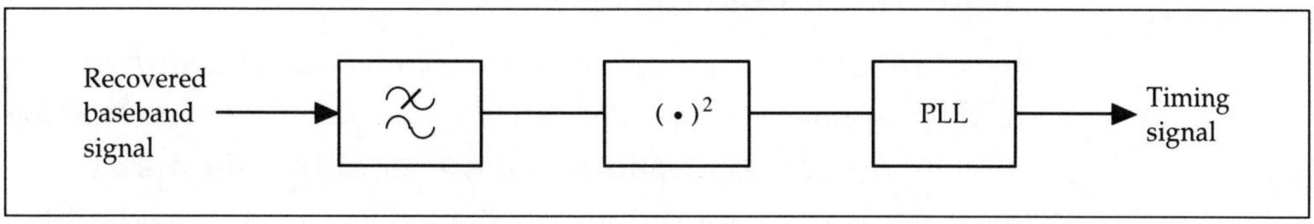

Figure E.5

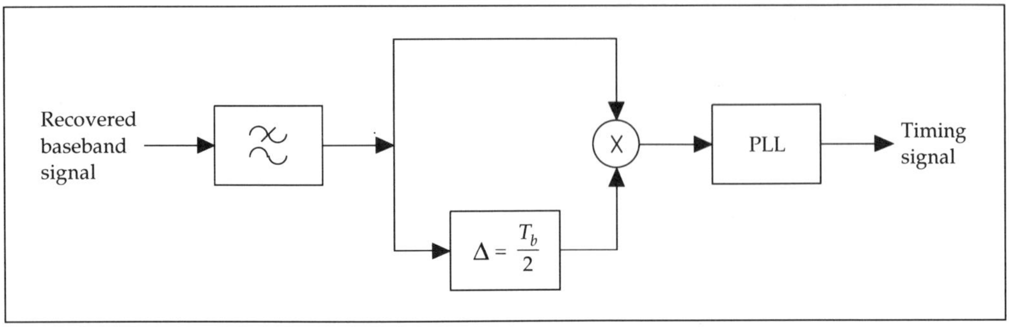

Figure E.6

2. Delay and Multiplier Technique

The demodulated baseband signal is multiplied by a delayed version of itself—the output of the multiplier contains a periodic component at the bit rate, which can be extracted by a PLL tuned at the frequency equivalent to the symbol rate. See Figure E.6.

Spectral Spreading of QPSK

Figure E.7 shows the effect of nonlinear phase and amplitude distortions on a QPSK signal. The large phase transitions of a QPSK signal (90°, 180°) when band limited have large envelope variations (up to 100%). When band-limited QPSK passes through a nonlinear amplifier working at saturation, the previously filtered side lobes are almost totally regenerated. To overcome this problem and maintain spectral efficiency additional processing can be employed.

MSK Schemes and Off-Set QPSK

In off-set QPSK, the quadrature stream is delayed with respect to the in-phase stream. This means that simultaneous changes of state do not occur and hence carrier amplitude variation is reduced. If the data pulses are shaped with a sinusoidal filter then minimum shift keying (MSK) results. In MSK, the rectangular-shaped pulses are converted to half sinusoidal pulses. See Figure E.8.

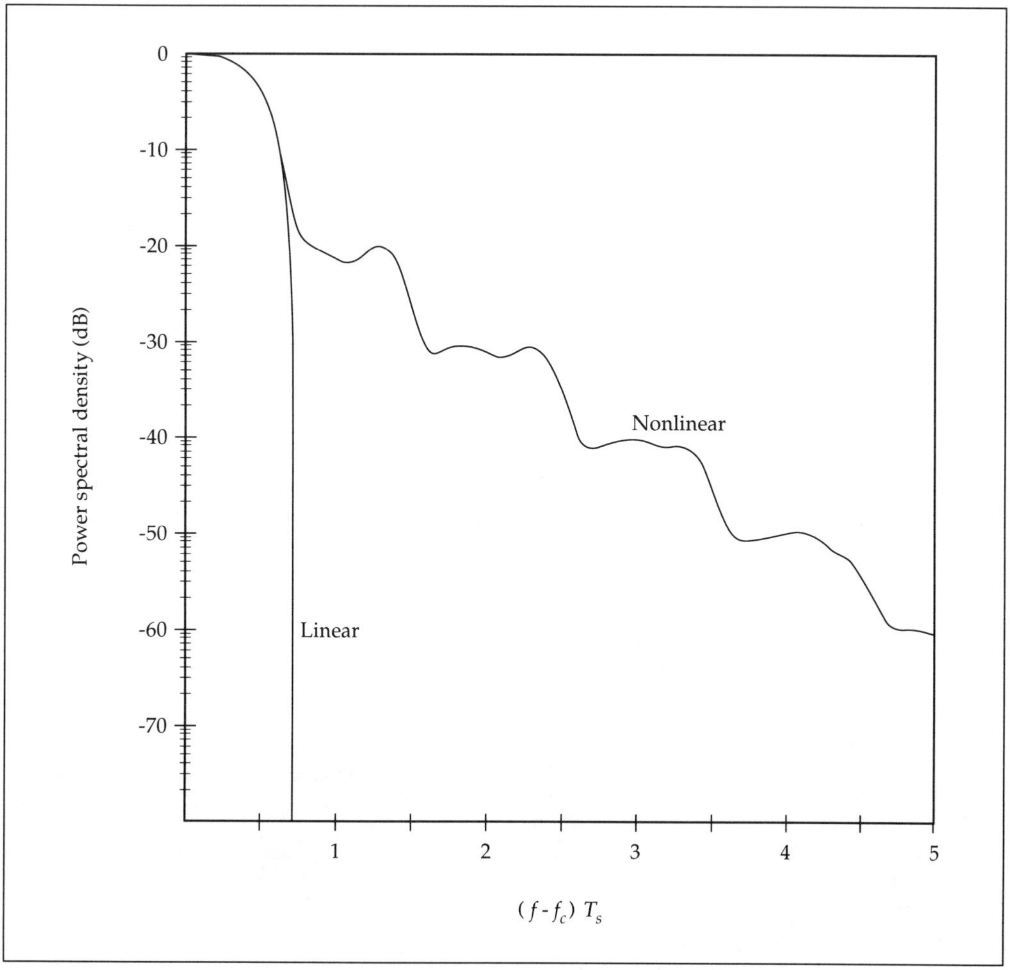

Figure E.7

Comparing QPSK, off-set QPSK, and MSK, the power spectral densities of QPSK and off-set QPSK are the same. The MSK main lobe is 1.5 times wider than QPSK but falls off at a higher rate. See Figure E.9.

Gaussian Minimum Shift Keying

In GMSK, as used in GSM, changes of 180° are not permitted. Varying the phase by ± 90° for each data bit allows the phasor to trace out a circle and maintain a constant envelope, avoiding

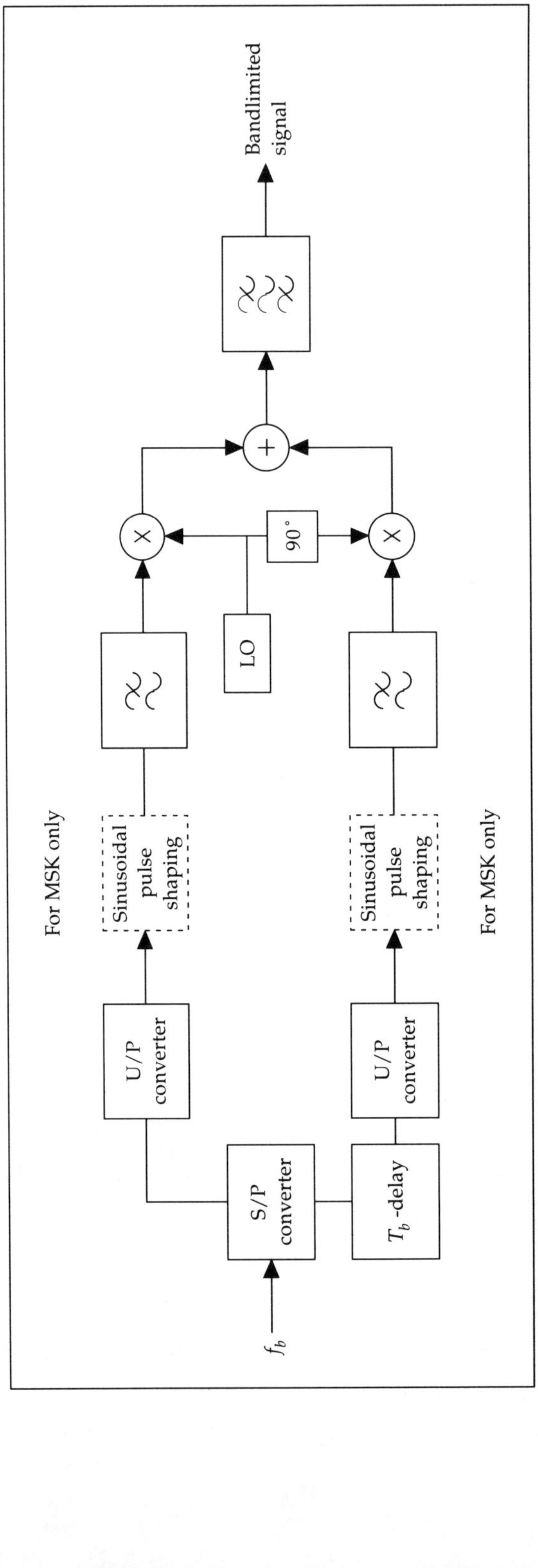
f_b
S/P converter
T_b-delay
U/P converter
U/P converter
Sinusoidal pulse shaping
Sinusoidal pulse shaping
For MSK only
For MSK only
LO
90°
Bandlimited signal

Figure E.8

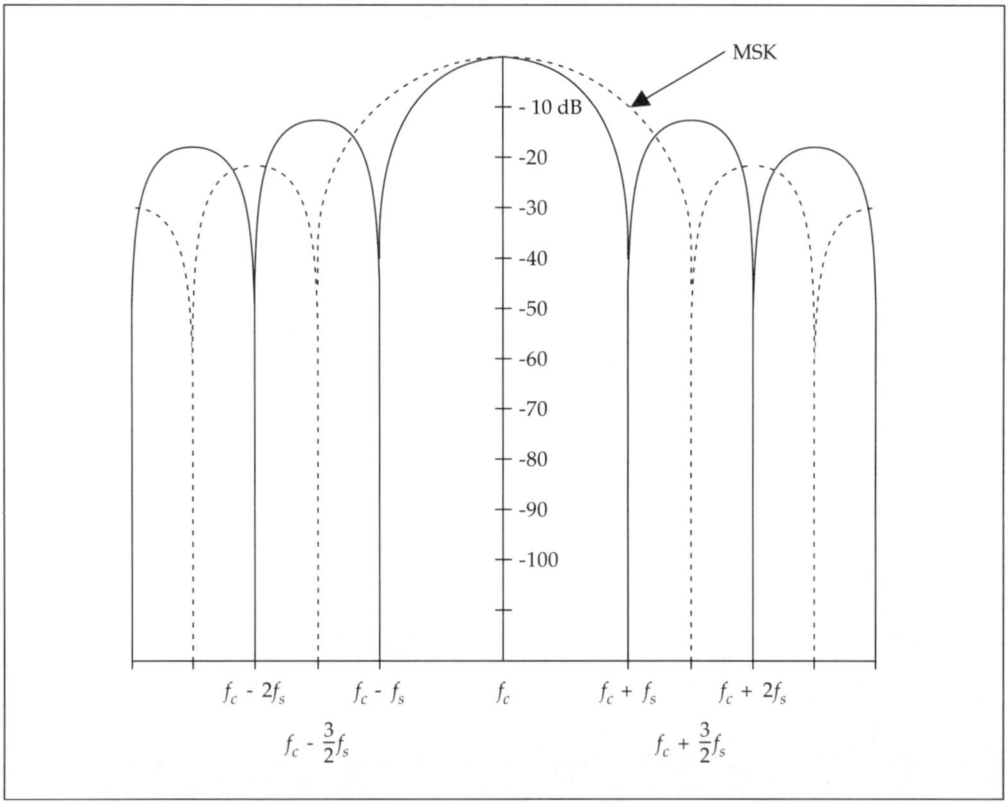

Figure E.9

adjacent channel interference (also sometimes described as "AM splash"). See Figure E.10. For random data, the average carrier frequency stays at the channel center. Figure E.11 shows a GMSK PLL-type modulator.

NOTE True QPSK implies an instantaneous change of phase at the symbol boundaries. Figure E.12 compares the modulation schemes.

QPSK with abrupt phase transitions (with just some bandwidth limiting) is compared to MSK (where the phase changes at a constant rate).

MSK has less sideband energy but maintains full eye opening at the symbol decision points.

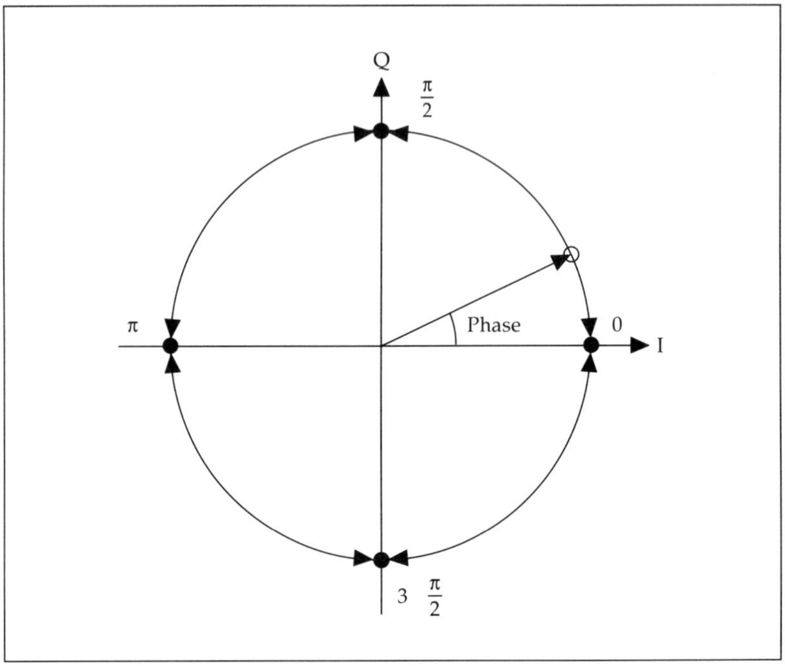

Figure E.10

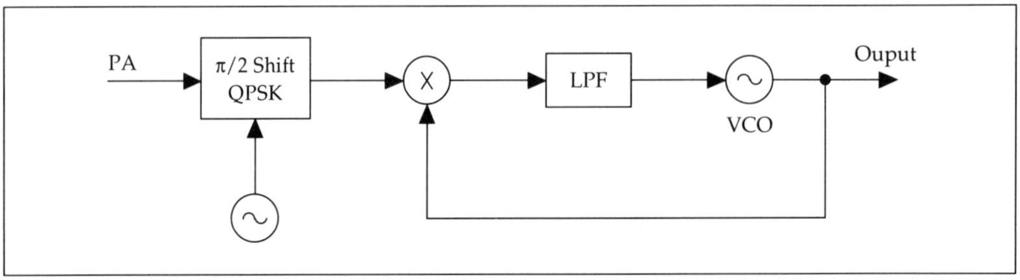

Figure E.11

GMSK has far less abrupt phase changes but the decision points are not always obtained. GSM accepts a maximum eye closure of 60% to achieve –60 dBc in adjacent channels.

The efficiency of modulation schemes is measured in terms of a 3 dB bandwidth/time ratio—compare for example modulation index for FM. In GMSK, as used in GSM, the BT is 0.3. This BT ratio is chosen to place the minimum energy in adjacent channels. Figure E.13 illustrates the trade-off in terms of signal degradation.

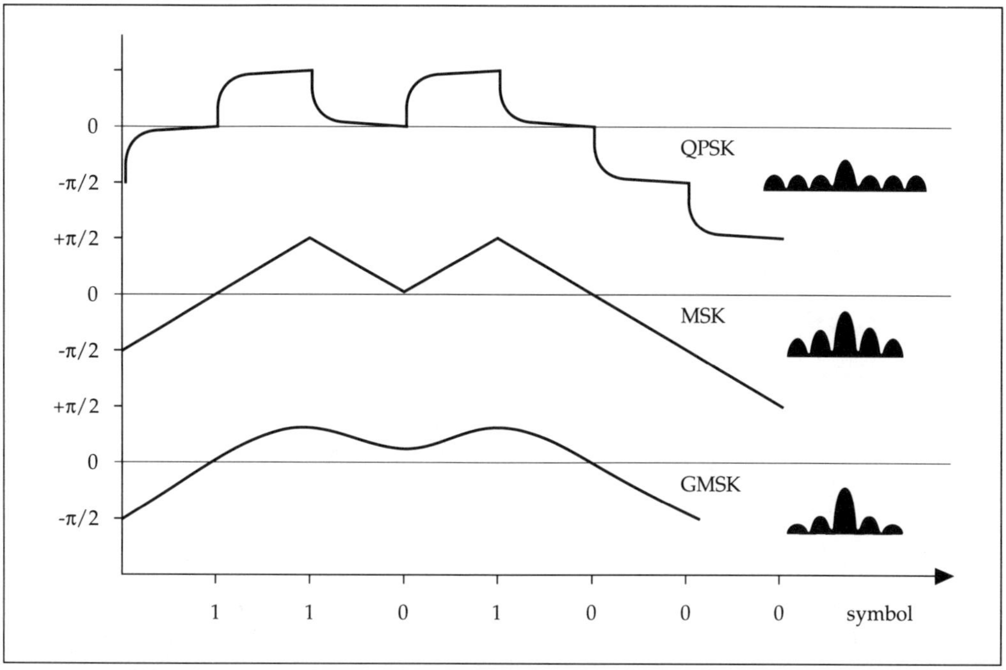

Figure E.12

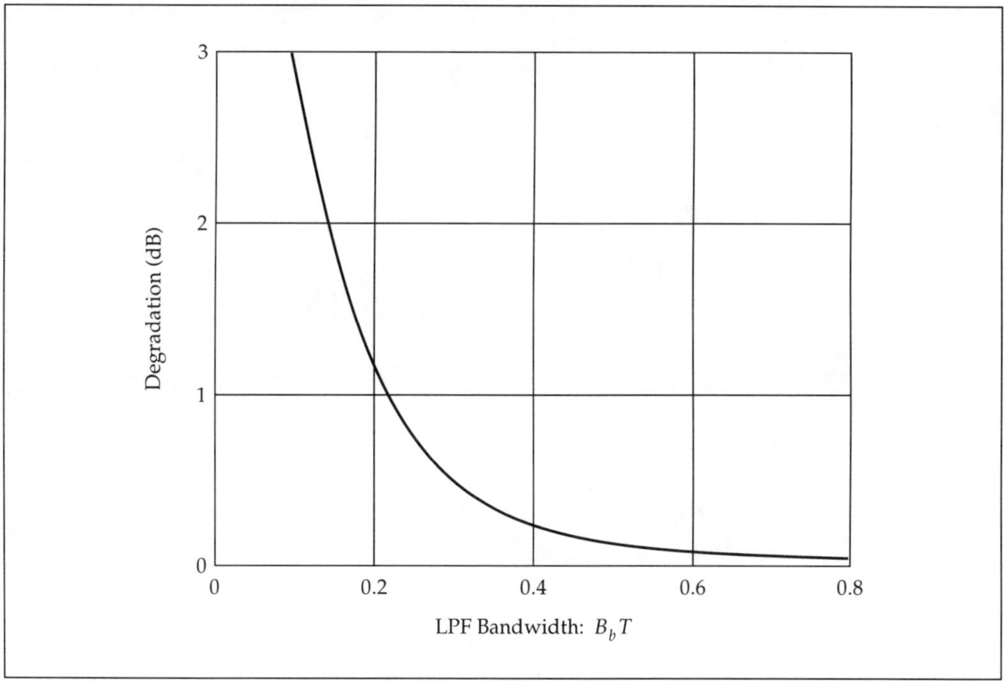

Figure E.13

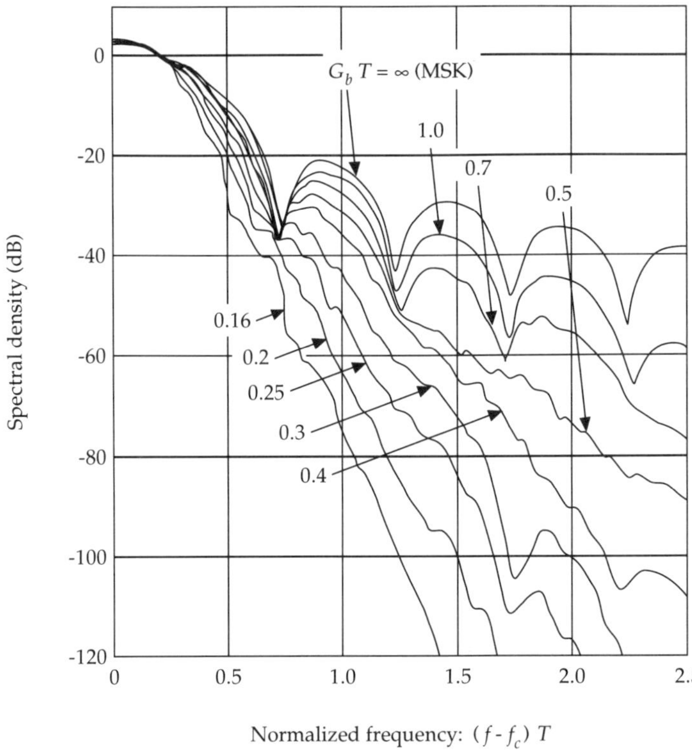

Figure E.14

The relative power spectral densities of GMSK at relative BTs is shown in Figure E.14.

Two typical examples of orthogonal coherent detection for MSK and GMSK are shown in Figure E.15.

π4 DQPSK

π4 DQPSK has been described briefly in our section on American Digital Cellular (ADC) (Chapter 9).

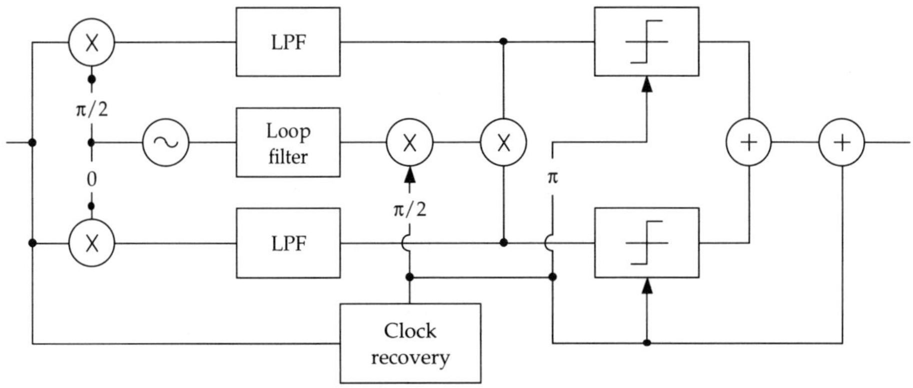

(a) Analog type

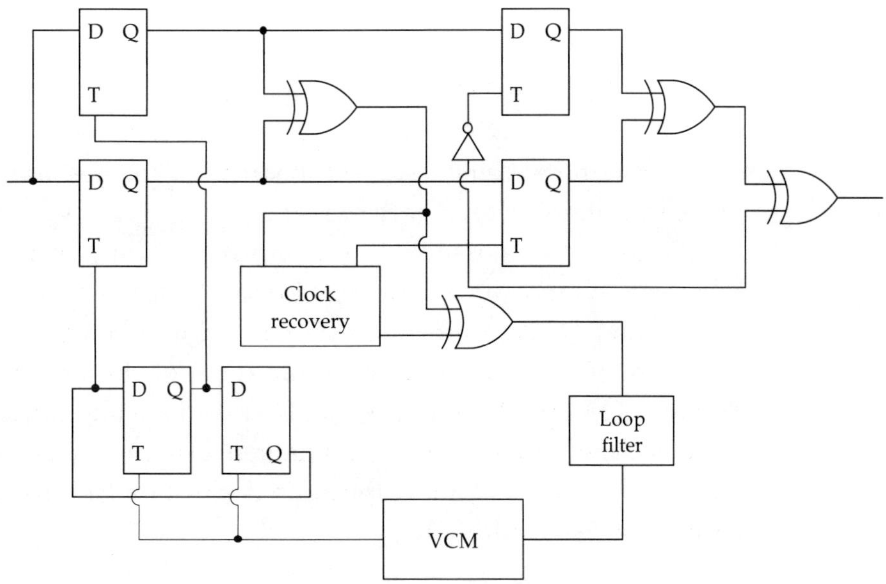

(b) Digital type

Figure E.15

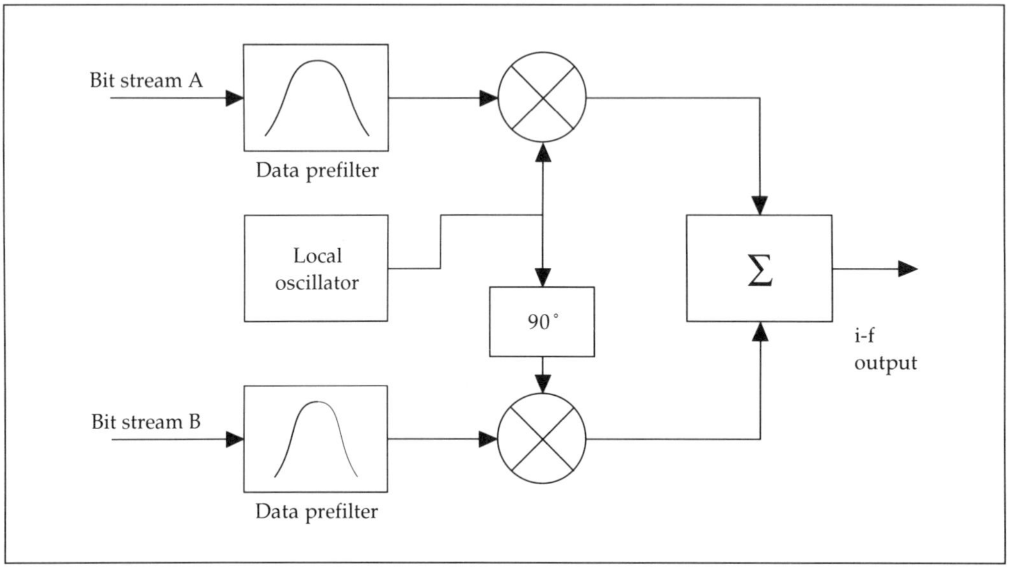

Figure E.16

π4DQPSK contains an AM component (i.e., envelope variation) equivalent to 70% depth, requiring transmitter amplifiers to be very linear. Being a 4-level modulation method implies phase transitions of greater than ± 90°.

To avoid the possibility of a zero phase shift when two successive symbols are the same, a 45° shift of axes occurs with each symbol. This results in a maximum shift of ± 135° and a minimum of ±45°, guaranteeing a transition for clock recovery. To limit AM splash and side bands spreading into adjacent channels, the applied data is prefiltered using a square root raised cosine filter to achieve the desired –60 dBc adjacent channel criteria. See Figure E.16.

Modulation Accuracy

Modulation accuracy for π4 DQPSK is defined as the distance of the phase vector from the calculated ideal. Typically the limit

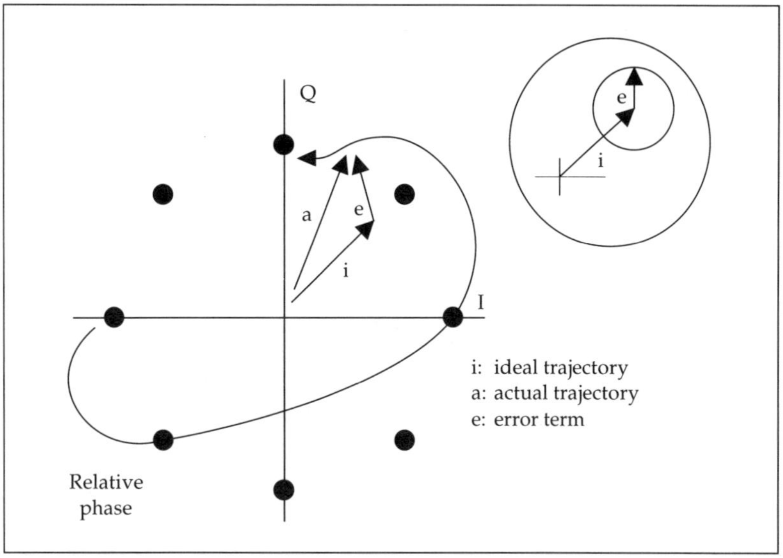

Figure E.17

will be ±12.5%, being the root sum of the squares of the phase and amplitude error. See Figure E.17.

In GSM, because of the constant amplitude, the error vector would be resolved into phase error and amplitude accuracy.

Spectrally Efficient Modulation Methods

A modulation technique is defined as spectrally efficient if it has a spectral efficiency of more than 2 bits/s/Hz.

Generally, it is easier to implement spectrally efficient, i.e., "higher level" modulation methods in fixed-link applications where the transmission path conditions are relatively stable. For example, 8-level PSK, 16-level QAM, 64-level QAM, 256 QAM or 1024 QAM. Higher level modulation methods in fixed-link applications result in lower costs per channel.

Higher level modulation methods are however progressively less resilient to system impairments and propagation constraints.

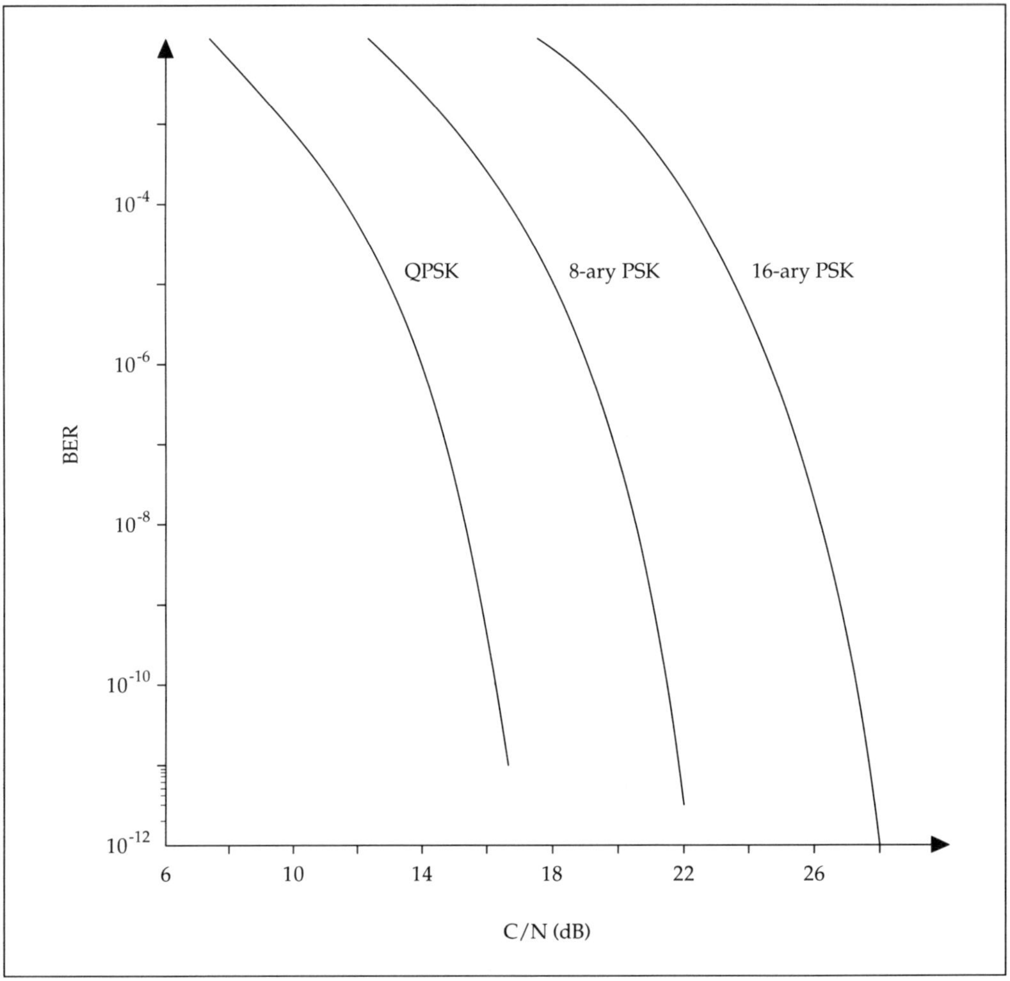

Figure E.18

8-level PSK and 16-level PSK are compared in Figure E.18 with QPSK in terms of their relative performance of bit error rate versus carrier-to-noise in dB.

The constellation of a 256-level QAM signal is shown in Figure E.19 with the related eye diagram (Figure E.20).

Table E.1 compares the theoretical spectral efficiency of the higher level modulation methods.

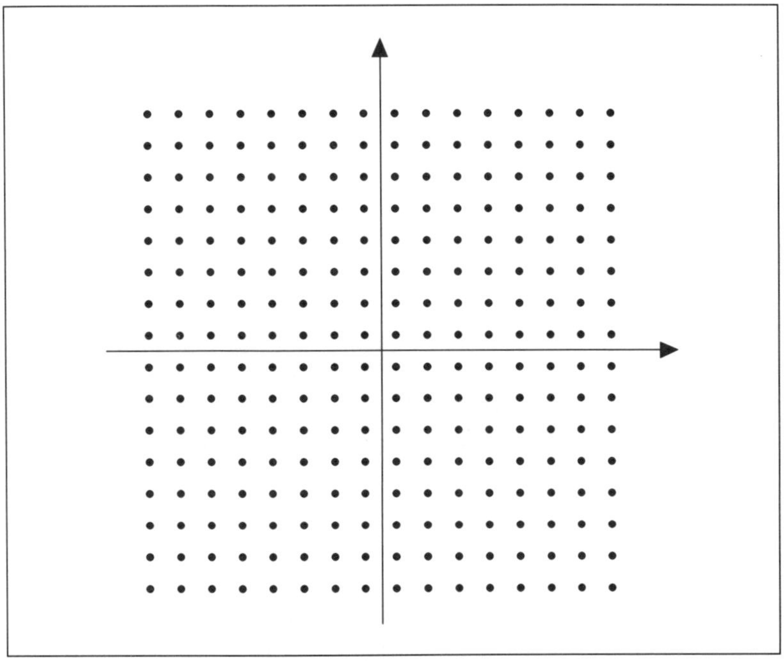

Figure E.19

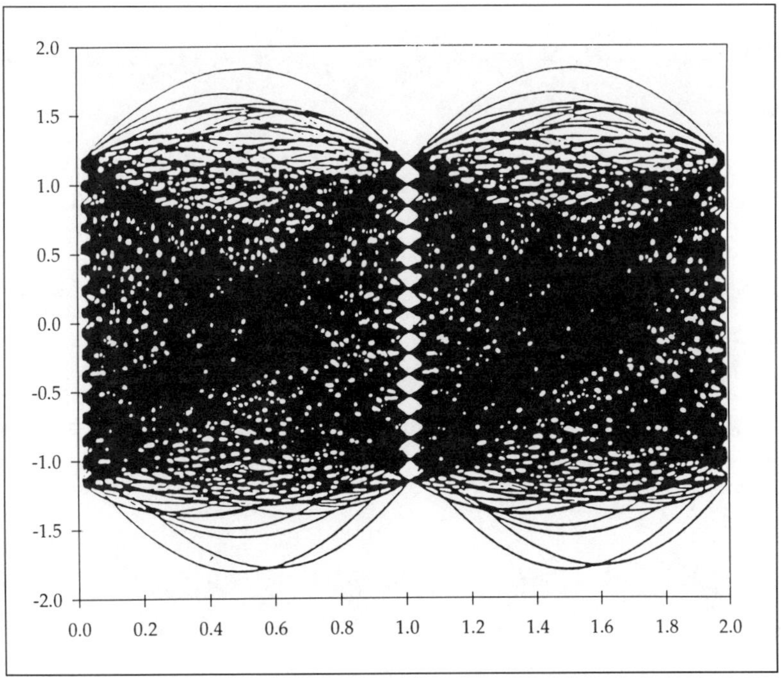

Figure E.20

Table E.1

M-ary QAM	**No. of levels in baseband signals**	**No. of bits per signaling element**	**Theoretical spectral efficiency (b/s/Hz)**
M = 16	4	4	4
M = 64	8	6	6
M = 256	16	8	8
M = 1024	32	10	10

In practice, 4-level modulation proves to be the best trade-off for most mobile radio applications.

APPENDIX F

HIGHER LEVEL MODULATION METHODS

We have avoided detailed coverage of higher level modulation methods such as 16 QAM, 64 QAM, 256 QAM, or QPRS options because of their limited relevance to mobile radio applications.

The reader is referred to the excellent reference work *Digital Communications*, written by Bernard Sklar and published by Prentice Hall.

APPENDIX

G

ISDN INTEGRATION

(Integrated Services Digital Network)

Increasing attention is being paid to the linking of mobile radio networks to the ISDN operating environment. The reader is referred to our chapter on GSM for an appreciation of the rate adaption and protocol adaption processes required in the translation gateways between the mobile and fixed operating environments.

This is an area in which standards log jams and technology log jams can occur. The reader is referred in particular to the section within the GSM chapter on delay and echo cancellation.

APPENDIX H

SOURCES OF UNITED KINGDOM PRODUCT DATA

We would like to thank the following for their help in providing product information.

Brian Winter
Spectronics Micro Systems Limited
Viking Way
Bar Hill
Cambridge CB3 8EL UK
(Tel: +44 954 780888)

Liz Baker
Paknet Limited
Coombe House, Coombe Square
Thatcham
Berkshire RG13 4FJ UK
(Tel: +44 635 72311)

John Haine
Cognito Group Limited
Medway House
Newbury Business Park
London Road
Newbury
Berks RG13 2PZ UK
(Tel: +44 635 508200)

David Pearce
Racal Vodata
1 The Pentangle
Park Street
Newbury
Berkshire RG13 1EA UK
(Tel: +44 635-33251)

Tim Shepherd
Dowty Systems Integration
Marshfield
Chippenham
Wiltshire SN14 8N UK
(Tel: +44 225 891881)

Steve Andrews
Orbitel
Keytech Centre
Ashwood Way
Basingstoke
Berks RG23 8BG UK
(Tel: +44 256 843468)

GLOSSARY

ACA adaptive channel allocation

AGC automatic gain control

ACT advanced CMOS

A–D analog-to-digital

ADC Americal digital cellular

ADPCM adaptive differential pulse code modulation

AF audio frequency

ARQ automatic repeat request

AMPS Precursor of TACS. Advanced mobile phone service; analog cellular standard of the United States and many "rest of the world" (RoW) countries.

ASCII American standard code for information interchange

ASIC application-specific integrated circuit

ASK amplitude shift keying

BAACH broadcast control channel

Band III A segment of spectrum (192–208 MHz) vacated by the discontinuation of black and white television broadcasting in the U.K. and colonized by two national U.K. and several regional 12½-kHz trunked networks.

Baseband The processing environment prior to up-conversion to i-f or rf. Typically a few bits per second (telemetry) to several hundred kilobits for digital cellular.

BCD binary coded decimal

BCH Bose Chaudhuri-Hocquenghem

BER bit error rate

bps bits per second

BPSK binary phase-shift keying

BSC base station controller

BTS base transceiver site

C centigrade

CAI common air interface

CDLC cellular data link control

CDMA code division multiple access

CELP codebook excitation linear prediction

churn "Soft" churn: user migrates from one carrier to another; "hard" churn: user discontinues service

C/I ratio carrier-to-interference ratio

CLAN cordless local area network

CLI calling line identification

CMOS complementary metal oxide semiconductor

cm^3 cubic centimeter

CT2 cordless telephone 2

CTCSS continuously controlled sub-audio squelch

CW continuous wave

D–A digital-to-analog

DAC digital-to-analog converter

dB decibels

dBm decibels relative to 1 milliwatt in 50 ohms

dc direct current

DCS digital controlled squelch

DDI direct digital interpolation

DDS direct digital synthesis

DECT Digital European Cordless Telephone System

DPSK differential phase shift keying

DRX discontinuous reception

DSI digitial speech interpolation

DSP digital signal processing

DSSS direct-sequence spread spectrum

DTI Department of Trade and Industry

DTX discontinuous transmission

E_b/N_o energy per bit to noise ratio

ECL emitter-coupled logic

EMC electromagnetic compatibility

EMMI electronic man–machine interface

EPOS electronic point of sale

ETACS extended total access communications system

E-TDMA extended time division multiple access

ETSI European Telecommunications Standards Institute

FAACH fast-associated control channel

FDM frequency division multiplexed

FDMA frequency division multiple access

FEC forward error correction

FFSK fast frequency shift keying

FMFB FM feedback

FSK frequency shift keying

fT transition frequency

g gram

GaAs gallium arsenide

GMSK Gaussian minimum shift keying

GP guard period

GPS global satellite positioning system

GSM Global System Mobile (formerly Groupe Speciale Mobile)

GDT group delay time

HCMOS high-speed complementary metal oxide semiconductor

HCT TTL-compatible HCMOS

HEMT high-electron mobility transistor

HF high frequency

Hz hertz

IC integrated circuit

i-f intermediate frequency (IF in CRH)

I^2L current injection logic

IN intelligent network

ISDN integrated services digital network

ISI intersymbol interference

JDC Japanese digital cellular

JTACS Japanese TACS

kb/s kilobits per second

kbps " " "

kg kilogram

kHz kilohertz

km/h kilometers per hour

LAN local area network

LO local oscillator

LSI large-scale integration

m meter

mA milliampere

MAHO mobile-assisted handover

MBE multiband excitation

mb/s megabits per second

MCN micro-cellular network

MESFET metal semiconductor field-effect transistor

MFSK minimum frequency shift keying

MHz megahertz

MIPS million instructions per second

mm millimeter

um micron (micrometer)

MMIC monolithic microwave integrated circuit

MPT Ministry of Post and Telecommunication

m/s meters per second

ms millisecond

MSA metropolitan service area

MSC mobile switch center

MSK minimum shift keying

MTBF mean time between failures

MTS mobile transceiver station

N-AMPS narrow-band amps

NCO numerically controlled oscillator

N-doped negatively doped semiconductor material

NDO numerically driven oscillator

NMT Nordic Mobile Telephone. Scandinavian analog cellular standard; implemented very successfully in many overseas countries.

N-TACS narrow-band TACS

OMC operations and maintenance center

OOK on-off keying

OSI open standards interconnection

PA power amplification

PCM pulse code modulation

PCN personal communication network

PLL phase-locked loop

PM phase modulation

PMR private mobile radio

PN pseudo random

POCSAG Post Office Standardization Advisory Group

PROM programmable-read-only memory

PSK phase shift keying

PSTN public switched telephone network

QPSK quaternary phase-shift keying

rads radians

RAM random-access memory

RELP-LTP residually excited linear prediction-long-termprediction

rf radio frequency

RLP radio link protocol

RoW rest-of-the-world

RS Reed Solomon

SAACH slow-associated control channel

SAT Supervisory audio tones. Three tones (5970, 6000, and 6030 Hz) used in TACS and AMPS analog cellular systems to detect co-channel interference.

SAW surface acoustic wave

SF stealing flags

SIM subscriber identity module

SMR specialized mobile radio

SSB single side band

ST Signalling tone at 8 kHz used for "clear down" of a call in analog cellular TACS or AMPS

TACS Total access communications system. Analog cellular standard "son of AMPS"—used in the U.K., Italy, Japan, and elsewhere.

TAR total accounting rate

TB tail bits

TDD time division duplex

TDMA time division multiple access

TIA Telecommunications Industry Association

TTE talk-time effect

TTL transistor–transistor logic

UMTS universal mobile telephone service

V volt

VLSI very large-scale integration

V-MACS Vodafone-managed access conversion service

VSAT very small aperture terminal

VSELP vector sum excited linear prediction

Wh watt-hour

ABOUT THE AUTHORS

Geoff Varrall graduated from St. John's College, Cambridge in the mid-1970s becoming a well-known marketing and technology specialist, developing a number of highly successful seminars in the electrical and electronic industries. More recently, he has acted as course principal for all RTT UK/European mobile radio technology programs.

Roger Belcher has a background of 25 years in the RF communications industry, including an apprenticeship at Marconi, design and development of antennae and radar processing equipment for Plessy Radar (1967–72), technical marketing for Texas Instruments (1972–77), antennae and communications for

Racal (1977–80), and design of RF test equipment and associated hardware/software development for Rohde and Schwartz (1980–84). He has been technical director of RTT since 1985.

Together, Varrall and Belcher provide technology assessment support to a broad client base within the mobile radio industry, including chip foundries, device manufacturers, carriers, service providers, and large multiple users; and to related industry sectors, telecommunications, broadcasting, the computer industry, the automotive majors, and the consultancy profession in the U.K., Europe, the United States, and Asia-Pacific.

INDEX